TITANIUM ALLOYS - ADVANCES IN PROPERTIES CONTROL

Edited by **Jan Sieniawski**
and **Waldemar Ziaja**

Titanium Alloys - Advances in Properties Control
http://dx.doi.org/10.5772/49999
Edited by Jan Sieniawski and Waldemar Ziaja

Contributors

James Wang, Farid Umarov, Yaroslav Matychak, Iryna Pohrelyuk, Viktor Fedirko, Oleh Tkachuk, Faramarz Djavanroodi, Wilson Wang, Poh Chye Khoon, Jan Sieniawski

Published by InTech

Janeza Trdine 9, 51000 Rijeka, Croatia

Copyright © 2013 InTech

Notice

Publishing Process Manager Viktorija Zgela
Technical Editor InTech DTP team
Cover InTech Design team

Additional hard copies can be obtained from orders@intechopen.com

Titanium Alloys - Advances in Properties Control , Edited by Jan Sieniawski and Waldemar Ziaja
p. cm.
ISBN 978-953-51-1110-8

For EU product safety concerns :
IN TECH d.o.o., Prolaz Marije Krucifikse Kozulić 3, 51000 Rijeka, Croatia,
info@intechopen.com or visit our website at intechopen.com.

Meet the editors

Professor Jan Sieniawski graduated in Materials Science from Silesian University of Technology in Gliwice in 1972. He has been engaged with the Faculty of Mechanical Engineering and Aeronautics of the Rzeszów University of Technology since 1974. He leads the Department of Material Science since 1996. He is one of the originators and prime mover of the R&D Laboratory for Aerospace Materials which he heads since 2004. Scientific activity of Prof. Sieniawski covers topics in the area of physical metallurgy of aerospace materials – titanium and aluminium alloys, nickel and cobalt-based superalloys, development and characterization of single crystal castings and thermal barrier coatings on the elements of the hot section of gas turbines. Prof. Sieniawski is a founder member of the Polish Materials Society, member of The Committee on Materials Science and Committee on Machine Building of the Polish Academy of Sciences. He holds the position of Editor-in-Chief of the quarterly journal of The Committee on Machine Building of the Polish Academy of Sciences "Advances of Manufacturing Science and Technology" and acts as a division editor of the scientific technical monthly journal "Mechanik". He is a member of international editorial board of the "Archives of Materials Science and Engineering" journal. Professor Sieniawski is an author and co-author of six books and over 300 papers in domestic and international scientific journals and conference proceedings.

Waldemar Ziaja graduated in Aeronautical Engineering from Rzeszów University of Technology in 1992 and ever since he is engaged with Department of Material Science at the Faculty of Mechanical Engineering and Aeronautics of the Rzeszów University of Technology. He obtained his PhD degree in 1999. Scientific interests of Dr. Ziaja include relationships between processing, microstructure and mechanical properties of the advanced structural materials, mainly titanium and nickel based alloys. He is also working on modelling the microstructure dependant deformation of materials in fatigue and creep conditions including damage initiation and growth. Dr. Ziaja is an author and co-author of over 50 papers in domestic and international scientific journals and conference proceedings.

Contents

Preface

Constant progress in engineering in the area of advanced structures and technologies for aerospace and power industry, medicine, automatics and mechatronics, manufacturing of control and measurement equipment, is driven to a large extent by development and application of new or modified functional materials. It includes titanium and its alloys, although initially they were intended mainly for use as a structural material for load, carrying airframe elements of aircraft and helicopters.

Titanium enables implementation of new solutions in structures and technologies due to its unique combination of physical, chemical and mechanical properties. First of all, it has a low density, high strength properties and good corrosion resistance in many chemical environments. Titanium is also characterized by low thermal conductivity and thermal expansion, high friction coefficient and propensity for seizure.

During the last decades, extensive research efforts have been concentrated on the evolution and control of the titanium alloys microstructure through adjustment of the processing parameters in order to obtain desirable balance of properties for specific applications.

Simultaneously, there was significant development of the methods for obtaining better properties of the surface layer of the elements made of titanium and its alloys. The aim was to increase hardness, prevent seizure in case of machine kinematic pairs and also to increase biocompatibility and corrosion resistance of medical implants in human body. The results of the studies falling into both subject areas are presented in this publication.

The book contains six chapters and covers topics dealing with biomedical applications of titanium alloys, surface treatment, relationships between microstructure and mechanical and technological properties, and the effect of radiation on the structure of the titanium alloys.

In the first chapter application of titanium alloys in orthopaedics is explored with emphasis on strategies for enhancing their bioactivity and promoting osseointegration of the implants.

The second chapter deals with techniques for surface modification of medical implants leading to their prolonged use. The results of the studies on the influence of sputtered hydroxyapatite and SiO_2 nanocoatings on titanium are discussed.

In the third chapter the processes of diffusive saturation of titanium by interstitial elements are modeled taking into account interaction of titanium with nitrogen and oxygen. The potentials of the model for estimating the effects of thermo-chemical treatment are discussed.

In the fourth chapter the relationships between processing parameters, morphology of the microstructure characterized by quantitative stereological parameters of microstructure and mechanical properties of selected high strength two-phase titanium alloys are summarized.

In the next chapter the formability of titanium alloys is discussed on the example of Ti-6Al-4V alloy. Forming limit diagrams are determined experimentally in the hydroforming bulge test and numerically, using finite element method and various constitutive models.

In the final chapter the influence of the radiation and others factors like plastic deformation and thermo-chemical treatment on the structure and crystal lattice defects in binary titanium alloys is investigated.

Prof. Jan Sieniawski
Faculty of Mechanical Engineering and Aeronautics
of the Rzeszów University of Technology

Chapter 1

Titanium Alloys in Orthopaedics

Wilson Wang and Chye Khoon Poh

Additional information is available at the end of the chapter

http://dx.doi.org/10.5772/55353

1. Introduction

Metallic implants are commonly used in the orthopedic field. Despite the large number of metallic medical devices in use today, they are predominantly make up of only a few metals. Metallic alloys such as titanium continue to be one of the most important components used in orthopaedic implant devices due to favorable properties of high strength, rigidity, fracture toughness and their reliable mechanical performance as replacement for hard tissues. Orthopaedic implants are medical devices used for the treatment of musculoskeletal diseases and may consist of a single type of biomaterial or comprise a number of different biomaterials working together in modular parts. Prime examples of titanium implants used in orthopaedics would include prosthetic hip and knee replacements for various types of arthritis affecting these joints, spinal fusion instruments for stabilizing degenerate and unstable vertebral segments, and fracture fixation devices of various types such as plates, screws and intramedullary rods. Although titanium based implants are typically expected to last ten years or more, however longevity is not assured and the lack of integration into the bone for long-term survival often occurs and leads to implant failure. Revision surgery to address such failure involves increased risk, complications and costs. The main reason for the failure of these implants is aseptic loosening which accounts for 60 to 70% of the cases for revision surgery. The success of implants is dependent on firm bonding or fixation of implant biomaterial to bone, for optimal function and lastingness. Therefore one of the key challenges in bone healing and regeneration is the engineering of an implant that incorporates osseointegration with enhanced bioactivity and improved implant-host interactions so as to reduce biological related implant failure.

1.1. Development of titanium alloys for use in orthopaedics

Titanium alloys, originally used for aeronautics, garnered attention from the biomedical field, due to their biocompatibility, low modulus of elasticity, and good corrosion resistance.

Nonetheless, it was the osseointegration phenomenon due to the presence of a naturally formed oxide layer on the titanium surface that sparked development of titanium for use in orthopaedics [1]. Titanium alloys are often used in non-weight-bearing surface components such as femoral necks and stems (Figure 1), as they have lower modulus of elasticity resulting in less stress shielding of bone [2]. Nonetheless the osseointegrative bioactivity is still often not sufficient to attain true adhesion between the implant and bone, which may ultimately lead to mechanical instability and implant failure [3].

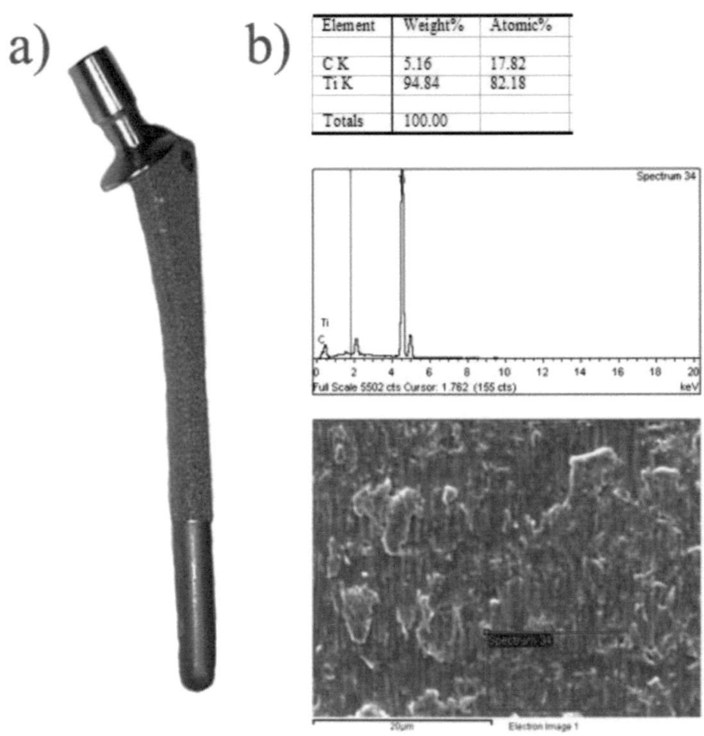

Figure 1. a) Titanium stem and (b) surface elemental analysis.

The mechanical properties of suitable titanium alloys based on Young's moduli should be similar to that of cortical bone. Cortical bone also termed compact bone is the major and most important constituent of the human skeleton and is crucial for bone functions including organ protection, movement, support etc. Young's moduli of β-type titanium alloys are substantially smaller than those of the α- and (α + β)-type alloys. This has brought on the discovery of harmless low-rigidity Ti alloys such as Ti-13Nb-13Zr, Ti-12Mo-6Zr-2Fe, Ti-15Mo-5Zr-3Al, Ti-15Mo, Ti-35Nb-7Zr-5Ta and Ti-29Nb-13Ta-4.6Zr. Nonetheless there are both advantages and disadvantages of the application of these titanium alloys. These alloys have proved to be effective in preventing bone atropy and enhancing bone remodeling, however the high amount of spring back and low fatigue strength make them undesirable as implant material. Ti-6Al-4V and commercial purity Ti are currently the most popular materials for implantation purposes. Commercial purity Ti has been tested to be inferior considering tensile strength, while Al and V have been shown to be unsafe. Currently researchers are still trying to develop other grades of alloys, such as Ti-6Al-7Nb and Ti–15Sn–4Nb–2Ta–0.2Pd. The most Ti alloys researched upon are the (α + β)-type alloys for their strength and ductility.

1.2. Bioactivity of titanium alloys

Each manufacturer of titanium implants has his own differing theories on implant designs for specific orthopaedic applications. Generally there are certain guiding principles that will affect the ultimate viability of an implant. The design of the implant has to take into account biomechanical and biological factors that may affect its success. Conformity to native anatomy, material properties and mechanical strength appropriate for the targeted function and environment are some of the considerations that come into play. Despite the benefits and successes of these medical devices, their use is not without risk of adverse effects. Titanium implants generally develop an oxide layer which allows it to integrate with living bone tissue. However, the body can have adverse reactions to titanium like fibrosis and inflammation which may affect its long term functional performance [4]. Success in the application of an orthopaedic implant would depend on various factors and implants may fail due to physiologic reasons such as aseptic loosening.

Generally there are two types of implant-tissue responses [5-7]. The first type is the response of the hosts' tissues to the toxicity of the implanted material. Implanted material may be toxic or release chemicals that could damage the surrounding tissues. The second response which is also the most common is the formation of a nonadherent fibrous capsule between the implant and the hosts' tissues termed fibrosis. This is a natural response to protect the body from a foreign object which may eventually lead to complete fibrous encapsulation [8]. Typically implants are intended to stay fixed in the human body for a long time and bone is expected to grow into the surface of the implant. Unfortunately this does not always happen. Fibrosis referred to as foreign body reaction, develops in response to almost all implanted biomaterials and consists of overlapping phases similar to those in wound healing and tissue repair processes [9]. Despite the biocompatibility of metallic implants used, titanium materials are generally encapsulated by fibrous tissue after implantation into the living body [10]. Cells trapped between the implant and the fibrous capsule also lack general housekeeping tissue functions like removing apoptotic or necrotic cells which can also promote chronic inflammation [11]. Not only that the ECM (extra cellular matrix) secreted by fibroblast is different from the bone matrix formation generated by osteoblast, in the long run this ECM layer may lead to micromotion and the generation of wear particles on the surfaces of the implant [12]. The resulting titanium debris may play a leading role in the initiation of the inflammatory cascade leading to osteolysis [13]. Eventually this causes aseptic loosening as the bonds of the implant to the bone are destroyed by the body's attempts to digest the wear particles. When this occurs the prosthesis becomes loose and the patient may experience instability and pain. Revision surgery to resolve this would entail further costs and morbidities to the patients. For bone tissue, direct osteoblast attachment on metal is important to prevent aseptic loosening of the metal implant caused by fibroblast layer attachment. Fibrosis can also cause osteoclast-independent bone resorption by fibroblast-like cells. It has been shown that fibroblast-like cells, under pathological conditions, not only enhance but also actively contribute to bone resorption [14]. Successful implant integration into the surrounding tissue is highly dependent on the crucial role of native cells, chiefly osteoblast attaching to the implant surface. Therefore one of the key challenges in orthopaedics is the engineering of an implant with enhanced osseointegration properties to reduce implant failure rates.

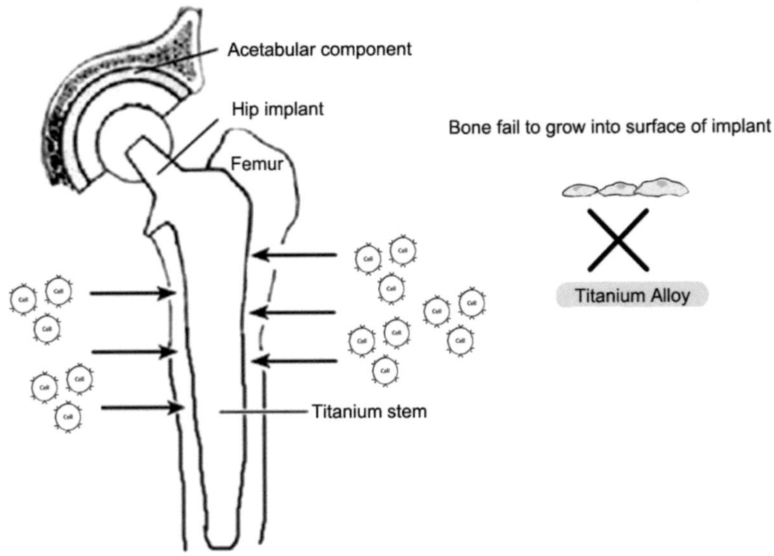

Figure 2. Schematic figure of a hip implant. The femoral neck is the region at risk of compromised vascularity. Arrows indicate area of compromised vascularity where osseiointegration fails to take place.

2. Strategies for conferring enhanced bioactivity to titanium alloys

So far most research efforts have been concentrated on improving the bone-implant interface, with the aim of enhancing bone healing and implant integration via either physical or chemical approaches [15]. The physical approach is focused on the modification of the implant surface morphology and topography using mechanical methods such as machining, acid-etching, plasma spraying, grit-blasting and anodization to improve the microtopography of the surface. The rationale behind this is that an increase in surface roughness of the implant material would provide a higher level of surface energy which would improve bone anchorage, matrix protein adsorption, osteoblasts functions and ultimately osseointegration [16].

The chemical approach is towards the creation of a bioactive implant surface via application of coatings onto the implant layer by biochemical and physicochemical techniques. In biochemical techniques, organic molecules such as growth factors, peptides or enzymes are incorporated to the implant layer to affect specific cellular responses [17]. While in physico-chemical techniques, the incorporation is achieved with inorganic phases such as calcium phosphate which may increase the biochemical interlocking between bone matrix proteins and surface materials thereby enhancing bone-bonding [16]. Many implant modifications may combine both physical and chemical engineering methods. In the following sections we will discuss some of the more popular strategies used to enhance implant integration and bone-bonding.

2.1. Inorganic coatings

Calcium phosphate coating has been widely used in the orthopaedic field due to their similarity with the mineral phase of bone [18] and are known for their bioactive properties which are beneficial in bone-bonding [19]. As calcium phosphate generally lacks the mechanical strength for use as bulk materials under loading conditions, they are often coated onto the surface of metallic implants. There are several studies published which have shown the favorable use of calcium phosphate coatings in increasing the biocompatibility of bone-implant interface, implant anchorage and integration [20]. The calcium phosphate layer functions as a physiological transition between the implant surface and the hosts' tissues which guides bone formation along the implant surface and the surrounding tissues. One of the most successful method for the application of calcium phosphate coatings is via the plasma-spraying method due to its advantage of extensive coating capability and high deposition rate. However despite numerous findings [21] that report the beneficial osteoinductive properties of plasma-sprayed calcium phosphate coatings, there are still some concerns regarding its use. Plasma-sprayed coatings are not uniform and there is poor control over thickness and surface topography, which may result in implant inflammation when particles are released from these heterogeneous coatings. To overcome these drawbacks, various other deposition strategies have been developed and employed such as biomimetic, electrophoretic and electrospray deposition etc. However care should be taken when comparing the efficacy of each of these methods which would require a comprehensive evaluation of both biological response and clinical performance. Although calcium phosphate coatings have been shown to be beneficial in enhancing bone-bonding, there is still no general consensus on the use of calcium phosphate coating systems. The main problems include large variation in the quality of calcium phosphate coatings, even between different batches and market forces which offer other cheaper alternatives [22].

2.2. Organic coatings

Surface modification of implant materials with growth factors and peptides is gaining popularity in the recent years [23, 24]. Various therapeutic biomolecules of interest can be immobilized onto implant surfaces to enhance the bone-implant interface interactions. Currently more popular approaches would include the immobilization of bone growth factors such as bone morphogenetic proteins (BMPs) to enhance osteogenesis and the deposition of peptide sequences to induce specific cellular functions. Growth factors immobilized on orthopaedic devices have been reported to enhance osteoblastic activity and favor implant integration [25]. The most commonly used growth factors in orthopaedics are members of the transforming growth factor beta (TGF-β) superfamily including the BMP family, especially BMP2 and BMP7. Growth factors may be physically adsorbed or covalently grafted onto the implant surface and various studies have shown that the loading of implant with these factors can enhance interactions at the bone-implant interface and aid the remodeling process ultimately improving implant integration [26-28]. However critical factors in the successful use of growth factors in orthopaedic devices are the optimum dosage, exposure period and release kinetics, all have to be considered carefully to avoid the detrimental effects associated with growth factor use such as high initial burst rate, ectopic bone formation and short half-life. More recently, peptide sequences with the ability to target specific osteogenic cellular functions of differentiation and mineralization have been developed [29, 30]. These short functional fragments derived from the original protein have increased shelf life, can be synthetically produced and are more resistant to denaturizing effects. Their usage would provide significant clinical benefits over the use of conventional proteins. They can be linked to the implant surface to provide biological cues for bone formation. Additionally other peptide sequences in use include the RGD, YIGSR, IKVAV and KRSR which have been used to improve cellular adhesion and bone matrix formation [31-33].

2.3. Organic–inorganic composite coatings

Research in the recent years have concentrated on the development of bioactive composite coatings which mimics the structure of the bone tissue. These composite coatings would combine calcium phosphate with growth factors, peptides, antibodies etc. to enhance interactions at the bone-implant interface. However due to the fact that often high temperature or non-physiological conditions are needed in the preparation of calcium phosphate coatings, only physical adsorption is employed in deposition of the biomolecules on the implant surface [34, 35]. However with physical adsorption techniques, initial high burst rate is often observed, which is not desired [36]. Therefore coating techniques that create a gentle sustained release kinetics are preferred. A recently published paper have shown that calcium phosphate coating combining slow release of antibiotics, aids in early success at recruitment of bone cells [37]. Many other studies have shown that depositing BMP2 and TGF-β onto the implant surface would greatly enhance bone-bonding at the bone-implant interface [25, 34]. The biological efficacy of orthopaedic implants can be improved greatly by both physical and chemical modifications. The use of a wide multitude of engineering techniques in the manipulation of surface topography, morphology and incorporating the use of various inorganic and organic components would directly influence the response in the local bone-implant interface and the apposition of new bone. With the development of new techniques and strategies on composite coatings to better mimic the human bone structure this would result in a new generation of orthopaedic implants with improved implant integration and bone healing.

3. Osseointegration of the implants

The clinical strategies to manage musculoskeletal defects would center around three components: cells, structure and growth factors. For the design of implant materials, cells and proteins at the implant interface plays a critical role [38]. The utilization of biosignal proteins such as growth factors for development of bioactive implant materials holds great potential. Especially due to the scarcity of stem cells in the body, materials which regulates cellular functions such as adhesion, growth and differentiation are desired.

One of the most important process in determining the success of an orthopaedic implant is osseointegration. Osseointegration is defined as the formation of a direct structural and functional connection between the living bone and the surface of a implant [39, 40]. An implant is considered osseointegrated if there is no progressive relative movement between the implant and the bone it has direct contact with [40]. Under ideal conditions, implants could permanently become incorporated within the bone and persist under all normal conditions of loading, that is the two could not be separated without fracture. Vascularization which is the provision of blood supply is a critical component for the process of osseointegration. The differentiation of osteogenic cells is highly dependent on tissue vascularity and ossification is closely linked to the vascularization of differentiating tissue [40]. Therefore the success of tissue healing, regeneration and integration lies in the key process of revascularization which is crucial in improving the successful integration of implants [41, 42].

Bone healing around implants involves a cascade of cellular and biological events that take place at the bone-implant interface until finally the entire surface of the implant is covered by newly formed bone. This cascade of biological events is regulated by differentiation of cells stimulated by growth factors secreted at the bone-implant interface [40]. There has been considerable interest in modifying implant surfaces with growth factors to improve their cell functions and tissue integration capacity at the bone-implant interface. Enhanced cell functions and cell substrate interactions have been demonstrated with growth factors immobilized onto implant materials [26-28]. One of the more important growth factors for stimulating neovascularization (i.e. formation of new blood vessels) in target areas [43] would be angiogenic growth factors, crucial in improving the successful integration of implants both *in vitro* and *in vivo* [41, 42]. Of these angiogenic factors, vascular endothelial growth factor (VEGF) is the most potent and widely used key regulator of neovascularization [43, 44]. VEGF is a crucial factor in not only angiogenesis regulation but also in osteoblast [45] and osteoclast function [46-48] during bone repair. VEGF acts directly on osteoblasts, promoting cell functions such as proliferation, migration and differentiation [49, 50]. In addition, VEGF also indirectly affect osteoblasts via its influences on endothelial cells [51, 52]. VEGF is known to induce endothelial cells in surrounding tissues to migrate, proliferate and form tubular structures [53] and is an essential survival factor for endothelial cells [51] and new vessel formation [54]. Endothelial cells are needed to provide complex interactive communication networks in bone for gap junction communication with osteoblasts crucial to their formation from osteoprogenitors [55]. Furthermore VEGF stimulates endothelial cells in the production of beneficial bone forming factors acting on osteoblasts [50]. In all, the effects of VEGF on osteoblasts, osteoclasts and endothelial cells may synergistically act to enhance bone formation.

3.1. Fixation of titanium implants

The fixation of prosthetic components to the bone can be done with or without bone cement. In the cemented technique polymethylmethacrylate (PMMA) is used to "glue" the metal to the bone. In direct biological fixation, precise bone cuts are required to achieve maximum contact between metal and bone. The advantage of cement fixation is that the prosthetic components are instantly fixed, allowing movement immediately after surgery. However in the instances where revision surgery is required, it is extremely difficult to chip out all the cement during implant replacement. Cement fixation is usually employed on elderly patients over sixty-five where their bone stock is more osteoporotic with less likelihood of growing into the prosthesis and chances of revision is lower due to less demands on the implant and shorter remaining life expectancy compared to younger patients. Direct biological fixation is generally used for young patients due to better bone stock and ingrowth potential. The disadvantage of biological fixation is that it can take weeks or months to be fully complete during which weight bearing activity is restricted. However the final fixation achieved is more natural with complete incorporation of implant within the bone in ideal situations. Furthermore in case of young patients the chances for future revision surgery is higher and it would be easier to revise a cementless prosthesis without the need for cement removal. Another problem perceived was that cementless titanium stems have been reported to be more resistant to osteolysis and mechanical failure compared to similar cemented titanium stems [56]. The features of titanium that are detrimental to the cement environment seems to have no effects in the cementless environment and may in fact be beneficial leading to differences in performance of the two techniques. Therefore the enhancement of the bone implant interface especially in direct biological fixation with titanium implants would be extremely useful. This would greatly reduce the lag period in which osseointegration occurs between the prosthesis and the patient's bone.

3.2. Surface functionalization by growth factors immobilization

One promising way to incorporate growth factors usage with implant materials would be by surface functionalization of growth factors. Soluble growth factors work by binding with cognate receptors on cells to form complexes which would result in autophosphorylation of the cytoplasmic domains of the receptors and this phosphorylation activates intracellular signal transduction. The formed complexes are then aggregated and internalized into the cells by both clathrin-dependent and clathrin-independent mechanisms which leads to the recycling of the receptors for degradatory down-regulation [57]. Similarly immobilized growth factors work by forming complexes with the cell surface receptors, however the signal transduction is expected to last longer than soluble growth factors due to the inhibition of the internalization process. Multivalency is another important phenomenon responsible for this prolonged enhanced mitogenic effect. Multivalent ligands interact and bind avidly to multiple

http://dx.doi.org/10.5772/55353

surface cell receptors through several binding modes. This enhances the formation of ligand-receptor complexes which are critical for signal transduction and the multivalent ligands are able to stabilize and prevent lateral diffusion of the formed complexes leading to the prolonged effect. Figure 3 shows the interactions of cells with the different forms of growth factor and the enhanced mitogenic effects.

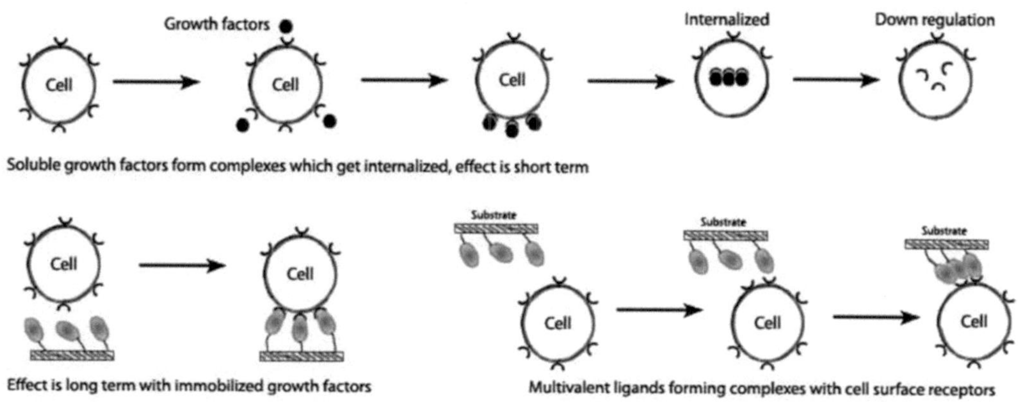

Figure 3. Effects of soluble growth factors compared to immobilized growth factors.

In order to effectively derive the effect from immobilized growth factors, strategies have to be developed that can optimize the structure to elicit the desired biological response. One of the problems encountered with implant materials for surface functionalizaton is the lack of suitable chemical groups on the surface. For more versatility and applicability, the concentrations of the OH group and other reactive groups such as amino or carboxyl groups have to be increased. The initial organic layer immobilized on the implant materials can then be used as a tether for biomolecular components used to mediate cell attachment. Another issue which merits investigation is the control of the retention and/or release of the biomolecules from the implant surface. The easiest and most common method employed for delivery of biomolecules is physical adsorption, which unfortunately provides little control over the delivery and orientation of the biomolecules. Bonding of the biomolecules and use of coatings incorporating them would be alternative methods of delivery to the bone-implant interface. Regardless, the preferred and chosen immobilization technique would depend on the specific working mechanism of the biomolecules. Given the above scenario, surface functionalization of biomaterials in order to enhance biocompatibility and promote osseointegration has great potential in addressing the problems of prosthetic joint implant longevity and survival.

Immobilization techniques are broadly classified into four categories, namely a) physical adsorption (via van der Waals or electrostatic interactions), b) physical entrapment (use of barrier systems), c) cross-linking and d) covalent binding. The choice of the technique would depend on the nature of the bioactive factors, substrates and its application. It will not be possible to have a universal means of immobilization, however developing a viable methodology which can provide for a facile, secure immobilization with good interactions for orthopaedic implants is vital.

3.2.1. Physical adsorption

This is the simplest of all the techniques available and does not alter the activity of the bioactive factors. Physical adsorption techniques are mainly based on ionic and hydrophobic interactions. If the bioactive factors are immobilized via ionic interactions, adsorption and desorption of the factors will depend on the basicity of the ion exchanger. A reversible dynamic equilibrium is achieved between the adsorbed factors and substrates which is affected by the pH as well as ionic strength of the surrounding medium. Hydrophobic interactions offer slightly higher stability with less loss of the factors from the surface of the substrates. Although physical adsorption systems are simple to perform and do not require extensive treatment to the bioactive factors and substrates used however there are certain drawbacks. These systems suffer from low surface loading and biomolecules may desorbed from the surface in an uncontrolled manner.

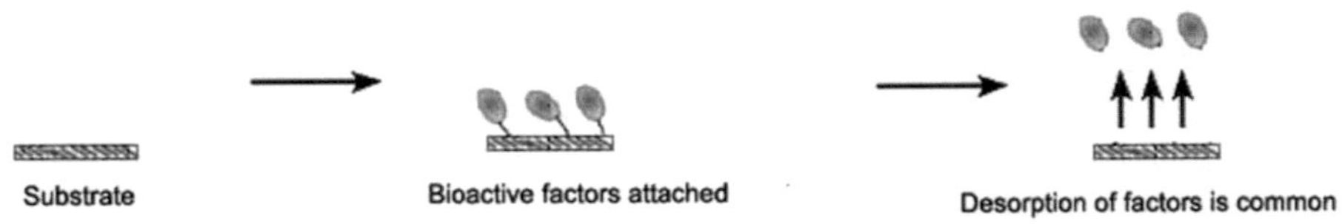

Figure 4. Schematic diagram showing physical adsorption system with proteins.

3.2.2. Physical entrapment

This method is employed with barriers including natural polymers like gelatin, agar and alginate entrapment systems. Other synthetic polymers employed include resins, polyurethane prepolymers etc. Some of the major limitations of the entrapment system is the diffusional problem where there is possible slow leakage during continuous use due to the small molecular size of bioactive factors, and steric hindrance which may affect the reactivity of the factors. Recent development of hydrogels and water soluble polymers attempt to overcome these drawbacks and have attracted much attention from the biomedical field.

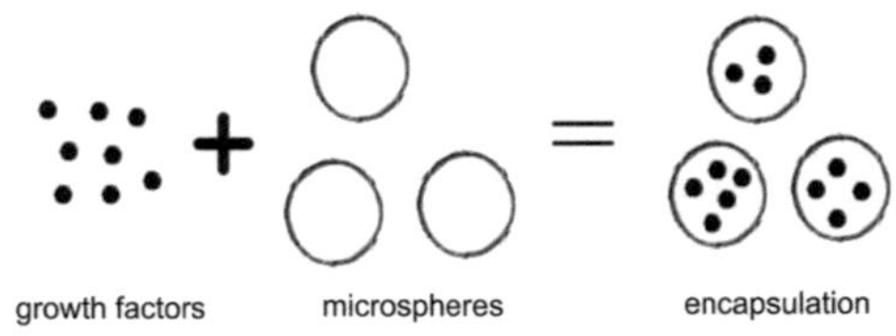

Figure 5. Schematic diagram showing barrier system with proteins.

3.2.3. *Cross-linking*

Bioactive factors can also be immobilized through chemical cross-linking via homo- as well as heterobifunctional cross-linking agents. Among these glutaraldehyde cross-linking are the most popular due to its low cost, high efficiency and stability [58-60]. Glutaraldehyde is often used as an amine reactive homobifunctional crosslinker for biochemistry applications.

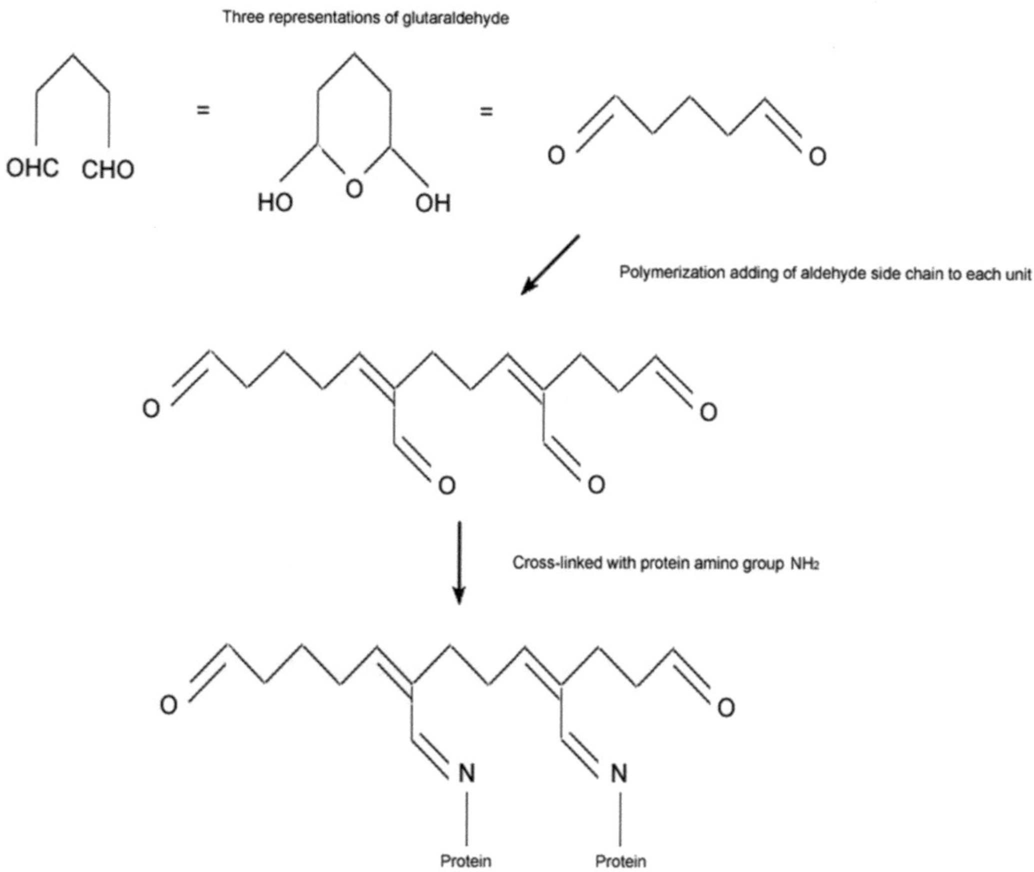

Figure 6. Schematic diagram showing glutaraldehyde cross-linking with proteins.

3.2.4. *Covalent binding*

Covalent binding is another technique used for the immobilization of bioactive molecules. The functional groups investigated are usually the carboxyl, amino and phenolic group of tyrosine. Bioactive factors are covalently linked through functional groups in the factors not essential for the bioactivity. The covalent binding should be optimized so as to protect the active site and not alter its conformational flexibility.

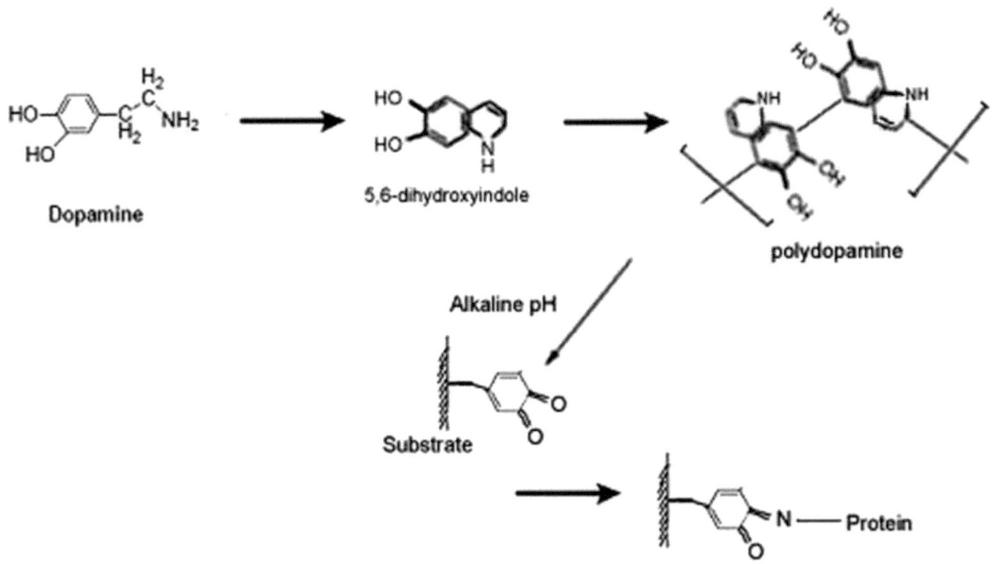

Figure 7. Schematic diagram showing polymerization of dopamine under alkaline pH and the equilibrium shift towards the quinone functional groups for reactivity with proteins.

3.2.5. Comparison of the various immobilization techniques

Several methods of immobilizing angiogenic growth factors onto substrates have been studied and reported [61-66]. A summary of a short study investigating the efficacy of immobilization of VEGF via various modes of functionalization on Ti-6Al-4V including physical adsorption, cross-linking and covalent binding (adapted for orthopaedic applications) is presented here to evaluate the effectiveness of each technique. As physical entrapment is not suitable in this case of improving the bone-implant interface via the surface of the implant material, therefore this system is not investigated. Table 1 summarizes the parameters of the binding efficiency, cytotoxicity, release profile and number of steps required for the fabrication of the substrates.

Although physical adsorption had the highest rate of binding however there was also uncontrolled release of the factors from the substrate which may be undesirable [67-69]. A measurement of the percentage of factors released into the solution over a 30 day period showed that more than 30% of the factors were released. A number of studies have examined simple coating or loading of factors onto implants [67-73] in order to provide local and sustained delivery after implantation. However with this strategy some studies showed an uncontrolled initial burst in the release kinetics of factors from such implants [67-69]. High levels of factors in the local microenvironments of these implants may be detrimental to healing and may promote tumorogenesis [74]. To avoid the deleterious effects, secure immobilization strategy would be preferred [61, 64-66]. Immobilization of growth factors on implants have been shown to promote desirable cell substrate interactions and enhance cell functions [62,

	Binding Efficiency (50ng loading)	Cytotoxicity	Factor release overtime	Active form	Number of steps required for fabrication
Physical adsorption (via simple coating)	86%	0.677	"/> 30% after 1 month	Soluble	Single step
Cross-linking (via glutaraldehyde cross-linking)	56%	0.449	Nil	Immobilized	Three steps
Covalent Binding (via polydopamine conjugation)	52%	0.841	Nil	Immobilized	Two steps

Table 1. please add caption

63]. Furthermore it has been demonstrated that immobilized factors is more effective in promoting proliferation of cells compared to soluble factors [65]. Both immobilized and soluble factors bind to receptors on cells, however they have differing effects due to the fact that soluble factors are internalized and subsequently degraded, while immobilization inhibits internalization and prevents down regulation [64, 75], thereby enabling the factors to stimulate proliferation for an extended period of time. A comparison of cross-linking and covalent binding shows that they come quite close in terms of binding efficiency and there is no release of growth factors into the solution which is the preferred methodology.

From the cytotoxicity indications (Table 1) follows that there is a lower cell viability with glutaraldehyde cross-linking compared to the other groups. This may be due to the fact that glutaraldehyde is known to be toxic and is able to kill cells quickly by cross-linking with their proteins. There have also been reports of its toxicity implicated in poor cell growth, attachment and apoptosis [58-60] by other groups. Although glutaraldehyde cross-linking effectively anchors a high density of factors onto the titanium substrate surface and the molecules are also more firmly attached than those which are physically adsorbed however the associated toxicity has made it unsuitable for clinical applications. The use of covalent immobilization with polydopamine looks promising. Polydopamine has been found to be able to form thin adherent films onto a wide variety of metallic substrates via covalent bonds and various strong intermolecular interactions including metal chelation, hydrogen bonding and π-π interactions [76] which cannot be disrupted by normal mechanical forces. The use of this bioreactive layer for covalent bioconjugation with bioactive factors for orthopaedic applications holds great potential. Although it will not be possible to have a universal means of immobilization, however it is vital to develop a viable methodology which can provide for secure immobilization with good interactions for orthopaedic implants. The choice of the technique would depend on the nature of the bioactive factors, substrates and their application. The development of surface modification procedures that do not affect the integrity of the substrate and bioactivity of the growth factors are crucial in producing the desired surface functionalization effect. This would provide us with a secure and efficient method of attaching bioactive molecules to titanium implant material surface conferring enhancement of cell-implant interactions beneficial for orthopaedic applications.

4. Conclusions

There is an ever growing need for orthopaedic advancement with the high prevalence and impact of musculoskeletal diseases. 50% of the world's population over 65 suffer from joint diseases and more than 25% of population over 65 require health care for joint related diseases. The instances for failed joint replacements associated with osteolysis and bone defects is increasing. There is an urgency to increase the success of bone implant fixation and the longevity of implant. Fixation of orthopaedic implants has been one of the most challenging and difficult problem faced by orthopaedic surgeons and patients. Fixation can often be achieved via direct biological fixation by allowing tissues to grow into the surfaces of the implants or with the use of bone cement acting as a grouting material. Whether cemented or cementless fixation are employed, the problems of micromotion and the generation of wear particles may eventually necessitate further surgery. Revision surgery poses increased risks like deep venous thrombosis, infection and dislocation, in addition to being an economic burden to the patient. Therefore the enhancement of implant integration would bring enormous benefits. Titanium alloy is one of the most frequently used material in orthopaedic implants. However despite the good inherent bioactivity and biocompatibility exhibited by titanium alloys, osseointegration with host tissue is still not definite, the lack of bioactivity may cause implant failure at times. Fixation of orthopaedic implants has been one of the most challenging and difficult problem faced by orthopaedic surgeons and patients. With the ever growing number of patients requiring orthopaedic reconstructions the development and evolvement of titanium alloys with structural and biological potential to manage bone healing impairment and defects would be desirable.

Author details

Wilson Wang* and Chye Khoon Poh

Department of Orthopaedic Surgery, National University of Singapore, Kent Ridge, Singapore

References

[1] Lausmaa J, Linder L. Surface spectroscopic characterization of titanium implants after separation from plastic-embedded tissue. Biomaterials. 1988;9:277-80.

[2] Niinomi M. Mechanical biocompatibilities of titanium alloys for biomedical applications. J Mech Behav Biomed Mater. 2008;1:30-42.

[3] Puleo DA, Nanci A. Understanding and controlling the bone-implant interface. Biomaterials. 1999;20:2311-21.

[4] Branemark R, Branemark PI, Rydevik B, Myers RR. Osseointegration in skeletal reconstruction and rehabilitation: A review. J Rehabil Res Dev. 2001;38:175-81.

[5] Mano T, Ueyama Y, Ishikawa K, Matsumura T, Suzuki K. Initial tissue response to a titanium implant coated with apatite at room temperature using a blast coating method. Biomaterials. 2002;23:1931-6.

[6] Rossi S, Tirri T, Paldan H, Kuntsi-Vaattovaara H, Tulamo R, Narhi T. Peri-implant tissue response to TiO2 surface modified implants. Clin Oral Implants Res. 2008;19:348-55.

[7] Migirov L, Kronenberg J, Volkov A. Local tissue response to cochlear implant device housings. Otol Neurotol. 2011;32:55-7.

[8] Suska F, Emanuelsson L, Johansson A, Tengvall P, Thomsen P. Fibrous capsule formation around titanium and copper. J Biomed Mater Res A. 2008;85:888-96.

[9] Kyriakides TR, Hartzel T, Huynh G, Bornstein P. Regulation of angiogenesis and matrix remodeling by localized, matrix-mediated antisense gene delivery. Mol Ther. 2001;3:842-9.

[10] Nebe JB, Muller L, Luthen F, Ewald A, Bergemann C, Conforto E, et al. Osteoblast response to biomimetically altered titanium surfaces. Acta Biomater. 2008;4:1985-95.

[11] Akbar M, Fraser AR, Graham GJ, Brewer JM, Grant MH. Acute inflammatory response to cobalt chromium orthopaedic wear debris in a rodent air-pouch model. J R Soc Interface. 2012;9:2109-19.

[12] Perona PG, Lawrence J, Paprosky WG, Patwardhan AG, Sartori M. Acetabular micromotion as a measure of initial implant stability in primary hip arthroplasty. An in vitro comparison of different methods of initial acetabular component fixation. J Arthroplasty. 1992;7:537-47.

[13] Kaufman AM, Alabre CI, Rubash HE, Shanbhag AS. Human macrophage response to UHMWPE, TiAlV, CoCr, and alumina particles: analysis of multiple cytokines using protein arrays. J Biomed Mater Res A. 2008;84:464-74.

[14] Pap T, Claus A, Ohtsu S, Hummel KM, Schwartz P, Drynda S, et al. Osteoclast-independent bone resorption by fibroblast-like cells. Arthritis Res Ther. 2003;5:R163-73.

[15] Castellani C, Lindtner RA, Hausbrandt P, Tschegg E, Stanzl-Tschegg SE, Zanoni G, et al. Bone-implant interface strength and osseointegration: Biodegradable magnesium alloy versus standard titanium control. Acta Biomater. 2011;7:432-40.

[16] Coelho PG, Granjeiro JM, Romanos GE, Suzuki M, Silva NR, Cardaropoli G, et al. Basic research methods and current trends of dental implant surfaces. J Biomed Mater Res B Appl Biomater. 2009;88:579-96.

[17] Morra M. Biomolecular modification of implant surfaces. Expert Rev Med Devices. 2007;4:361-72.

[18] Rey C. Calcium phosphate biomaterials and bone mineral. Differences in composition, structures and properties. Biomaterials. 1990;11:13-5.

[19] de Groot K, Wolke JG, Jansen JA. Calcium phosphate coatings for medical implants. Proc Inst Mech Eng H. 1998;212:137-47.

[20] Barrere F, van der Valk CM, Meijer G, Dalmeijer RA, de Groot K, Layrolle P. Osteointegration of biomimetic apatite coating applied onto dense and porous metal implants in femurs of goats. J Biomed Mater Res B Appl Biomater. 2003;67:655-65.

[21] Geesink RG. Osteoconductive coatings for total joint arthroplasty. Clin Orthop Relat Res. 2002:53-65.

[22] Wennerberg A, Albrektsson T. Structural influence from calcium phosphate coatings and its possible effect on enhanced bone integration. Acta Odontol Scand. 2009:1-8.

[23] van den Beucken JJ, Vos MR, Thune PC, Hayakawa T, Fukushima T, Okahata Y, et al. Fabrication, characterization, and biological assessment of multilayered DNA-coatings for biomaterial purposes. Biomaterials. 2006;27:691-701.

[24] Bierbaum S, Hempel U, Geissler U, Hanke T, Scharnweber D, Wenzel KW, et al. Modification of Ti6AL4V surfaces using collagen I, III, and fibronectin. II. Influence on osteoblast responses. J Biomed Mater Res A. 2003;67:431-8.

[25] Liu Y, de Groot K, Hunziker EB. BMP-2 liberated from biomimetic implant coatings induces and sustains direct ossification in an ectopic rat model. Bone. 2005;36:745-57.

[26] Hall J, Sorensen RG, Wozney JM, Wikesjo UM. Bone formation at rhBMP-2-coated titanium implants in the rat ectopic model. J Clin Periodontol. 2007;34:444-51.

[27] Schmidmaier G, Wildemann B, Cromme F, Kandziora F, Haas NP, Raschke M. Bone morphogenetic protein-2 coating of titanium implants increases biomechanical strength and accelerates bone remodeling in fracture treatment: a biomechanical and histological study in rats. Bone. 2002;30:816-22.

[28] Sumner DR, Turner TM, Urban RM, Turek T, Seeherman H, Wozney JM. Locally delivered rhBMP-2 enhances bone ingrowth and gap healing in a canine model. J Orthop Res. 2004;22:58-65.

[29] Saito A SY, Ogata S, Ohtsuki C, Tanihara M. Accelerated bone repair with the use of a synthetic BMP-2-derived peptide and bone-marrow stromal cells. J Biomed Mater Res A. 2005 Jan;72:77-82.

[30] Choi J-Y, Jung U-W, Kim C-S, Eom T-K, Kang E-J, Cho K-S, et al. The effects of newly formed synthetic peptide on bone regeneration in rat calvarial defects. J Periodontal Implant Sci. 2010;40:11-8.

[31] LeBaron RG, Athanasiou KA. Extracellular matrix cell adhesion peptides: functional applications in orthopedic materials. Tissue Eng. 2000;6:85-103.

[32] Ranieri JP, Bellamkonda R, Bekos EJ, Vargo TG, Gardella JA, Jr., Aebischer P. Neuronal cell attachment to fluorinated ethylene propylene films with covalently immobilized laminin oligopeptides YIGSR and IKVAV. II. J Biomed Mater Res. 1995;29:779-85.

[33] Schliephake H, Scharnweber D, Dard M, Rossler S, Sewing A, Meyer J, et al. Effect of RGD peptide coating of titanium implants on periimplant bone formation in the alveolar crest. An experimental pilot study in dogs. Clin Oral Implants Res. 2002;13:312-9.

[34] Alam MI, Asahina I, Ohmamiuda K, Takahashi K, Yokota S, Enomoto S. Evaluation of ceramics composed of different hydroxyapatite to tricalcium phosphate ratios as carriers for rhBMP-2. Biomaterials. 2001;22:1643-51.

[35] Lin M, Overgaard S, Glerup H, Soballe K, Bunger C. Transforming growth factor-beta1 adsorbed to tricalciumphosphate coated implants increases peri-implant bone remodeling. Biomaterials. 2001;22:189-93.

[36] Siebers MC, Walboomers XF, Leewenburgh SC, Wolke JC, Boerman OC, Jansen JA. Transforming growth factor-beta1 release from a porous electrostatic spray deposition-derived calcium phosphate coating. Tissue Eng. 2006;12:2449-56.

[37] Reiner T, Gotman I. Biomimetic calcium phosphate coating on Ti wires versus flat substrates: structure and mechanism of formation. J Mater Sci Mater Med. 2010;21:515-23.

[38] Hirano Y, Mooney DJ. Peptide and protein presenting materials for tissue engineering. Adv Mater. 2004;16:17-25.

[39] Novaes AB, Jr., de Souza SL, de Barros RR, Pereira KK, Iezzi G, Piattelli A. Influence of implant surfaces on osseointegration. Braz Dent J. 2010;21:471-81.

[40] Mavrogenis AF, Dimitriou R, Parvizi J, Babis GC. Biology of implant osseointegration. J Musculoskelet Neuronal Interact. 2009;9:61-71.

[41] Freed LE, Guilak F, Guo XE, Gray ML, Tranquillo R, Holmes JW, et al. Advanced tools for tissue engineering: scaffolds, bioreactors, and signaling. Tissue Eng. 2006;12:3285-305.

[42] Laschke MW, Harder Y, Amon M, Martin I, Farhadi J, Ring A, et al. Angiogenesis in tissue engineering: breathing life into constructed tissue substitutes. Tissue Eng. 2006;12:2093-104.

[43] Boontheekul T, Mooney DJ. Protein-based signaling systems in tissue engineering. Curr Opin Biotechnol. 2003;14:559-65.

[44] Ferrara N, Davis-Smyth T. The biology of vascular endothelial growth factor. Endocr Rev. 1997;18:4-25.

[45] Deckers MM, Karperien M, van der Bent C, Yamashita T, Papapoulos SE, Lowik CW. Expression of vascular endothelial growth factors and their receptors during osteoblast differentiation. Endocrinology. 2000;141:1667-74.

[46] Kaku M, Kohno S, Kawata T, Fujita I, Tokimasa C, Tsutsui K, et al. Effects of vascular endothelial growth factor on osteoclast induction during tooth movement in mice. J Dent Res. 2001;80:1880-3.

[47] Engsig MT, Chen QJ, Vu TH, Pedersen AC, Therkidsen B, Lund LR, et al. Matrix metalloproteinase 9 and vascular endothelial growth factor are essential for osteoclast recruitment into developing long bones. J Cell Biol. 2000;151:879-89.

[48] Nakagawa M, Kaneda T, Arakawa T, Morita S, Sato T, Yomada T, et al. Vascular endothelial growth factor (VEGF) directly enhances osteoclastic bone resorption and survival of mature osteoclasts. FEBS Lett. 2000;473:161-4.

[49] Mayr-wohlfart U, Waltenberger J, Hausser H, Kessler S, Gunther KP, Dehio C, et al. Vascular endothelial growth factor stimulates chemotactic migration of primary human osteoblasts. Bone. 2002;30:472-7.

[50] Midy V, Plouet J. Vasculotropin/vascular endothelial growth factor induces differentiation in cultured osteoblasts. Biochem Biophys Res Commun. 1994;199:380-6.

[51] Gerber HP, McMurtrey A, Kowalski J, Yan M, Keyt BA, Dixit V, et al. Vascular endothelial growth factor regulates endothelial cell survival through the phosphatidylinositol 3'-kinase/Akt signal transduction pathway. Requirement for Flk-1/KDR activation. J Biol Chem. 1998;273:30336-43.

[52] Roberts WG, Palade GE. Increased microvascular permeability and endothelial fenestration induced by vascular endothelial growth factor. J Cell Sci. 1995;108 (Pt 6): 2369-79.

[53] Soker S, Machado M, Atala A. Systems for therapeutic angiogenesis in tissue engineering. World J Urol. 2000;18:10-8.

[54] Alon T, Hemo I, Itin A, Pe'er J, Stone J, Keshet E. Vascular endothelial growth factor acts as a survival factor for newly formed retinal vessels and has implications for retinopathy of prematurity. Nat Med. 1995;1:1024-8.

[55] Villars F, Guillotin B, Amedee T, Dutoya S, Bordenave L, Bareille R, et al. Effect of HUVEC on human osteoprogenitor cell differentiation needs heterotypic gap junction communication. Am J Physiol Cell Physiol. 2002;282:C775-85.

[56] Emerson RH, Jr., Head WC, Emerson CB, Rosenfeldt W, Higgins LL. A comparison of cemented and cementless titanium femoral components used for primary total hip

arthroplasty: a radiographic and survivorship study. The Journal of arthroplasty. 2002;17:584-91.

[57] Ito Y. Covalently immobilized biosignal molecule materials for tissue engineering. The Royal Society of Chemistry 2008. 2007;4:46-56.

[58] Gough JE, Scotchford CA, Downes S. Cytotoxicity of glutaraldehyde crosslinked collagen/poly(vinyl alcohol) films is by the mechanism of apoptosis. J Biomed Mater Res. 2002;61:121-30.

[59] Lin FH, Yao CH, Sun JS, Liu HC, Huang CW. Biological effects and cytotoxicity of the composite composed by tricalcium phosphate and glutaraldehyde cross-linked gelatin. Biomaterials. 1998;19:905-17.

[60] Huang-Lee LL, Cheung DT, Nimni ME. Biochemical changes and cytotoxicity associated with the degradation of polymeric glutaraldehyde derived crosslinks. J Biomed Mater Res. 1990;24:1185-201.

[61] Hall H, Hubbell JA. Matrix-bound sixth Ig-like domain of cell adhesion molecule L1 acts as an angiogenic factor by ligating alphavbeta3-integrin and activating VEGF-R2. Microvasc Res. 2004;68:169-78.

[62] Ito Y, Liu SQ, Imanishi Y. Enhancement of cell growth on growth factor-immobilized polymer film. Biomaterials. 1991;12:449-53.

[63] Karakecili AG, Satriano C, Gumusderelioglu M, Marletta G. Enhancement of fibroblastic proliferation on chitosan surfaces by immobilized epidermal growth factor. Acta Biomater. 2008;4:989-96.

[64] Sharon JL, Puleo DA. Immobilization of glycoproteins, such as VEGF, on biodegradable substrates. Acta Biomater. 2008;4:1016-23.

[65] Shen YH, Shoichet MS, Radisic M. Vascular endothelial growth factor immobilized in collagen scaffold promotes penetration and proliferation of endothelial cells. Acta Biomater. 2008;4:477-89.

[66] Yao C, Prevel P, Koch S, Schenck P, Noah EM, Pallua N, et al. Modification of collagen matrices for enhancing angiogenesis. Cells Tissues Organs. 2004;178:189-96.

[67] Patil SD, Papadmitrakopoulos F, Burgess DJ. Concurrent delivery of dexamethasone and VEGF for localized inflammation control and angiogenesis. J Control Release. 2007;117:68-79.

[68] Elcin AE, Elcin YM. Localized angiogenesis induced by human vascular endothelial growth factor-activated PLGA sponge. Tissue Eng. 2006;12:959-68.

[69] Kempen DH, Lu L, Heijink A, Hefferan TE, Creemers LB, Maran A, et al. Effect of local sequential VEGF and BMP-2 delivery on ectopic and orthotopic bone regeneration. Biomaterials. 2009;30:2816-25.

[70] Lode A, Wolf-Brandstetter C, Reinstorf A, Bernhardt A, Konig U, Pompe W, et al. Calcium phosphate bone cements, functionalized with VEGF: Release kinetics and biological activity. Journal of Biomedical Materials Research - Part A. 2007;81:474-83.

[71] Lode A, Reinstorf A, Bernhardt A, Wolf-Brandstetter C, Konig U, Gelinsky M. Heparin modification of calcium phosphate bone cements for VEGF functionalization. Journal of Biomedical Materials Research - Part A. 2008;86:749-59.

[72] Matsusaki M, Sakaguchi H, Serizawa T, Akashi M. Controlled release of vascular endothelial growth factor from alginate hydrogels nano-coated with polyelectrolyte multilayer films. J Biomater Sci Polym Ed. 2007;18:775-83.

[73] Petersen W, Pufe T, Starke C, Fuchs T, Kopf S, Raschke M, et al. Locally applied angiogenic factors--a new therapeutic tool for meniscal repair. Ann Anat. 2005;187:509-19.

[74] Ehrbar M, Djonov VG, Schnell C, Tschanz SA, Martiny-Baron G, Schenk U, et al. Cell-demanded liberation of VEGF121 from fibrin implants induces local and controlled blood vessel growth. Circ Res. 2004;94:1124-32.

[75] Place ES, Evans ND, Stevens MM. Complexity in biomaterials for tissue engineering. Nat Mater. 2009;8:457-70.

[76] Lee H, Dellatore SM, Miller WM, Messersmith PB. Mussel-inspired surface chemistry for multifunctional coatings. Science. 2007;318:426-30.

Chapter 2

Sputtered Hydroxyapatite Nanocoatings on Novel Titanium Alloys for Biomedical Applications

Kun Mediaswanti, Cuie Wen, Elena P. Ivanova,
Christopher C. Berndt and James Wang

Additional information is available at the end of the chapter

http://dx.doi.org/10.5772/54263

1. Introduction

Titanium and titanium alloys have been extensively studied for many applications in the area of bone tissue engineering. It was believed that the excellent properties of titanium alloys, e.g. lightweight, excellent corrosion resistance, high mechanical strength and low elastic modulus compared to other metallic biomaterials such as stainless steels and Cr-Co alloys, would provide enhanced stability for load-bearing implants. However, they usually lack sufficient osseointegration for implant longevity, and their biocompatibility is also an important concern in these applications due to the potential adverse reactions of metallic ions with the surrounding tissues once these metallic ions are released from the implant surfaces. One approach for consideration to improve the healing process is the application of a hydroxyapatite nanocoating onto the surface of biomedical devices and implants. Hydroxyapatite, with its excellent biocompatibility, and similar chemistry and structure to the mineral component of bone, provides a bioactive surface for direct bone formation and apposition with adjacent hard tissues. The deposition of a SiO_2 interlayer between the implant surface and the hydroxyapatite nanocoating is necessary to further improve the biocompatibility of metal implants, as SiO_2 has its own excellent compatibility with living tissues, and high chemical inertness, which lead to enhanced osteointegrative and functional properties of the system as a whole.

Therefore, SiO_2 and hydroxyapatite nanocoatings were deposited onto titanium alloys using electron beam evaporation and magnetron sputtering techniques, respectively, with different process parameters to optimize the deposition conditions and so achieve desired properties. Surface characteristics are essential due to their role in enhancing osseointegration. Surface morphology and microstructure were observed using a scanning electron microscope (SEM) and elemental analysis was performed by the energy dispersive X-ray spectroscopy method (EDS). The crystal structure was examined using X-ray diffractometer (XRD) to identify the phase components, while nanocoating thickness was measured using profilometer.

This chapter is divided into five major parts. First is an overview of bone and bone implants, including their structure and mechanical properties. The second part highlights the importance of nanocoatings for bone implants longevity. Various coatings and surface modification techniques of titanium and its alloys are also elucidated. The advantages and drawbacks of each technique are reviewed. The last part focuses on the study of sputtered hydroxyapatite and SiO_2 nanocoatings on titanium. A thorough discussion of the results is presented.

2. Natural bone and bone implants

2.1. Natural bone

Bone is a complex living tissue that harnesses the synergies of osseous tissue, cartilage, dense connective tissues, epithelium, adipose tissue and nervous tissue. Bone as a functional organ in the human body has various roles, such as supporting soft tissues, protecting many internal organs, enabling movements in human activity and facilitating mineral homeostasis, *i.e.,* storage of osseous tissue minerals such as calcium and phosphate, providing blood cell production sites and acting as a location for triglyceride storage [1].

Bone consists of both organic and inorganic materials that are distributed within an extracellular matrix. Organic material, called fibrous protein collagen, is predominant in bone structure and this collagen contributes to the tensile strength of bone. The inorganic material impregnated inside bone is mainly hydroxyapatite, *i.e.,* minerals of calcium phosphate and calcium carbonate. Usually, the calcium to phosphorus ratio of natural bone ranges between 1.50-1.65 depending on its location. Around 25 wt.% of bone consists of water that is present in bone pores, thereby ensuring nutrient diffusion and contributing to the viscoelastic properties of the material. Calcification is a process of crystallisation of mineral salts *i.e.,* calcium phosphate, which occurs in the biological framework formed by the collagen fibres [2].

There are four types of cells in osseous tissues: osteogenic cells, osteoblasts, osteocytes and osteoclasts. Osteogenic cells undergo cell division and develop into osteoblasts. Osteoblasts play a role in bone formation and collagen secretion. As osteoblasts secrete extracellular matrix, then osteoblasts evolve into osteocytes. Osteocytes, also known as mature bone cells, are responsible for nutrients and waste exchange with the blood. Osteoclasts are bone destroying cells and responsible for bone resorption. Bone consists of bone lining cells, fibroblasts, and fibrocytes. Bone lining cells control the movement of ions between bone and the surrounding tissue. The role of fibroblasts and fibrocytes is, in brief, to form collagen [1].

Bone can be categorized into five types on the basis of its shape, namely long, short, flat, irregular, and sesamoid. In addition to the dense structures present, osseous tissue has many small spaces between its cellular and extracellular matrix. There are two types of osseous tissue on the basis of the size and distribution of these spaces: compact bone tissue and spongy bone tissue. About 80 wt.% of the human skeleton is compact bone tissue. Compact bone consists of a packed osteon within the Haversian architecture. Each osteon consists of a central Haversian canal, concentric lamellae, lacunae, osteocytes, and canaliculi. Spongy bone, also termed as trabecular bone, exhibits a porous structure with porosity ranging from 50-90 wt.% and consists of an integrate lamellae network. The role of trabeculae is to support and protect the red bone marrow [2].

Bone structure contains macro, micro and nanoscale pores with different functions and characteristics. Macro-scale porosity gives rise to mechanical anisotropy. Micro-scale porosity provides sufficient vascularisation and cell migration, while nanoscale features act as a framework for cell and mineral binding [2].

Bone mechanics is determined mainly by the bone structure. Compact bone is stiffer and stronger than cancellous bone. The mechanical properties of human bone are listed in Table 1 [2]. The elastic modulus of human bone is approximately 0.05-2 GPa for cancellous bone and 7-30 GPa for compact bone [2]. It should be kept in mind that "elastic modulus" is not an exact description for bone properties since they are anisotropic and viscoelastic.

Mechanical Properties	Human Haversian (MPa)
Tensile strength	158
Tensile yield stress	128
Compressive strength	213
Compressive yield stress	180
Shear strength	71

Table 1. Mechanical properties of human haversian [2]

2.2. Bone implant

The history of implants started with the applications of autograph, allograph, and artificial device techniques [3]. Autographs utilized tissues from other parts of the patient's body, whilst allograft techniques used tissue from a donor. However, both techniques had drawbacks in application. The autograph method was limited only to nose bone and finger junctions [3]. Moreover, there were adverse side effects, such as infections and pain at the implant area. The allograft technique required a compatible donor that matched the patient's body system, which was usually difficult to find. There was always the potential risk of infections and disease transmission from the donor to the recipient's body. Artificial grafts employed artificial materials, now known as biomaterials. The advantages of using artificial device grafts include (i) lower risk for any transmission of disease, (ii) a reduced risk of infections, and (iii) the availability of many biomaterials for potential use as scaffolds. Therefore, ongoing studies aim to develop a new generation of biomaterials for bone implants.

2.3. Criteria of ideal bone implant

An ideal bone implant material should be osteoconductive, osteoinductive and should have osseointegration ability [3]. Furthermore, other key criteria for excellent implant performance include biocompatibility and mechanical compatibility. In addition, any implant waste after degradation should not cause harmful effects to the body. Recent trends in bone tissue engineering studies have revealed that bone implants may also serve as a drug delivery system if they are appropriately designed.

Osteoconduction is a process by which bone is directed to conform to a material's surface, while osteoinduction is the ability of an implant to induce osteogenesis. An inductive agent will stimulate undifferentiated cells to form preosteoblasts [3]. According to Branemark *et al.* [4], osseointegration could be defined as the "continuing structural and functional co-existence, possibly in a symbiotic manner, between differentiated, adequately remodelled, biologic tissues, and strictly defined and controlled synthetic components, providing lasting, specific clinical functions without initiating rejection mechanisms".

In the context of orthopaedic implants, the development of a drug delivery system is still at an early developing stage. The promising concept of using an implant as part of a drug delivery system could be described as the integration of therapeutic agents and devices.

In addition to high mechanical strength, the Young's modulus is a critical mechanical property in an artificial device when designing materials for bone implants. Other fundamental requirements for an ideal orthopaedic biomedical implant include high wear resistance, good fatigue properties if used under cyclic loading, no adverse tissue reactions, and high corrosion resistance.

3. Titanium and titanium alloys as bone implant materials

The applications of titanium in modern society, such as aviation and military defence, have been exploited widely. Titanium components have also been used in biomedical devices, including screws, plates, and hip and knee prostheses, for either bone fractures or bone replacement. These proven applications can be attributed to the distinctive properties of titanium and its alloys; properties such as high strength to density ratio and high corrosion resistance that enable their use as bone substitutes under load bearing conditions. Moreover, titanium exhibits a high tensile strength that is not featured in polymer or ceramic biomaterials. However, the long term inertness of titanium towards human tissues after implantation is a major drawback, as this means a lack of direct chemical bonding between the implant and host tissues [5-6].

Another concern regarding the use of solid titanium is that the dense structure is unable to support new bone tissues in growth and vascularisation. In addition, titanium has a much higher elastic modulus than natural bone, *i.e.,* 5 GPa and 110 GPa for bone and dense titanium, respectively [7-8]. This biomechanical mismatch causes stress shielding and, eventually, may lead to aseptic loosening that results in additional surgery after 10-15 years of implantation [9]. The development of porous titanium may potentially overcome problems of this nature.

The development of new titanium alloys has been extensively explored. Usually Al, Sn, O, C, N, Ga, and Zr are used as α stabilizers, while V, Mo, Ta, Nb, and Cr are used as β stabilizers [10]. Titanium alloys such as Ti6Al4V with aluminium and vanadium as α and β stabilizing elements have been widely used as implant materials. These first generation biomedical titanium alloys, however, have revealed that the release of Al and V metal ions is harmful to the human body [11]. The decisive requirement of a biomedical implant is its biocompatibility in the human body. Thus, alloying elements must be carefully chosen to reduce any biologically adverse impacts. Alloying elements that attract biomedical applications are Ta, Nb, and Zr due to their non-cytotoxicity, good biocompatibility, high corrosion resistance and their complete solid solubility in titanium [10].

Beta alloys that have higher β stabilizers content are attracting great interest for bone implant applications due to their low elastic modulus. Beta alloys that have been studied for bone implant applications include Ti50Ta20Zr, Ti64Ta, Ti13Nb13Zr, Ti42Nb, and Ti30Zr10Nb10Ta. Studies conducted by Obbard *et al.* [12] showed that by adjusting the concentration of β stabilizer Ta, the elastic modulus could be reduced. In this fashion the compliance mismatch between the implant and bone would be reduced, leading to lesser stress shielding.

Alpha-beta alloys may have some advantages over β alloys, namely lower density and higher tensile ductility. Some studies have succeeded in the production of alpha-beta alloys with a porous structure. The porous structure serves as an anchorage for bone in-growth and exhibits a lower elastic modulus, while the α and β phases provide sufficient mechanical strength for load bearing applications.

The development of porous titanium alloys with a variety of alloy components has brought about many improvements in bio-mechanical properties. For example, porous Ti10Nb10Zr with 69% porosity exhibited a tensile strength of 67 MPa, while pure Ti and pure Ta scaffolds with the same porosity demonstrated lower strengths of 53 MPa and 35.2 MPa, respectively [13]. Xiong *et al.* [14] reported that the elastic moduli of porous Ti-26Nb alloys with porosity of 50, 60, 70, and 80% were 25.4, 11.0, 5.2 and 2.0 GPa, respectively, while the plateau strength ranged from 180 MPa to 11 MPa.

4. The importance of nano-coatings for bone implant materials

Surface modification is a process that changes the composition, microstructure and morphology of a surface layer while maintaining the mechanical properties of the material. The aim of surface modification is to improve the bioactivity of the biomaterials so that the biomaterials demonstrate a higher apatite inducing ability that, in turn, leads to rapid osseointegration. After surface treatment, it is expected that the implant's surface will form an active apatite layer. The role of the thin apatite layer is to be a bonding interface to stimulate bone apatite and collagen production [15-16]. It is suggested that altering the nanostructured surface morphology influences the apatite inducing ability and improves osteoblast adhesion and differentiation [17].

4.1. Calcium phosphate coatings

Calcium phosphate is a synthetic ceramic that has been proven to support bone apposition and to enhance the osteoconduction of the bone. Calcium phosphate ceramics for bone tissue applications include tricalcium phosphate (TCP), octocalcium phosphate (OCP), hydroxyapatite ($Ca_{10}(PO_4)_6(OH)_2$, HA), and biphasic calcium phosphate (BCP) [18]. These ceramics accelerate the healing process and have been widely used in conjunction with metallic material as a bioactive coating material. The ratio of Ca/P in calcium phosphate should resemble the biological apatite mineral of bone (*i.e.*, 1.50-1.69). Calcium phosphate has the natural facility to bond directly to bone.

4.2. Nano-hydroxyapatite coatings

Hydroxyapatite demonstrates the best bioactivity amongst all the forms of calcium phosphate. Hydroxyapatite (HA) exhibits functionality in promoting osteoblast adhesion, migration, differentiation and proliferation; all of which are essential for bone regeneration. HA also has the ability to bond directly onto bone. The bioactivity of HA has made this ceramic the favourite for implant applications. HA nanoparticles may also induce cancer cell apoptosis [19]. The crystalline form of HA exhibits biointegration and prevents formation of adverse fibrous tissue. It is a more desirable coating than amorphous HA due to its ability to provide a better substrate for a different cell line [20]. Amorphous HA tends to dissolve in human fluid more easily and leads to loosening of the implant. Nanocrystalline HA is more favourable than microcrystalline HA because of its structural similarity with apatite [21].

5. Surface modification techniques

5.1. Sol-gel

The sol-gel method has been widely used to deposit calcium phosphate onto dense or porous metallic materials. There are two routes for a sol-gel reaction, namely inorganic and organic, using reagents consisting of a colloidal suspension solution of inorganic or organic precursors. The sol-gel technique transforms a liquid (sol) into a solid phase (gel) and requires drying and heat treatment stages. The advantages of the sol-gel method include: (i) it is cost-effective, (ii) it is easy to control the final chemical composition and thickness of the coating, (iii) the coating is readily anchored on the substrate, and it is usually homogenous with a good surface finish, and (iv) it can be used for coating implants or substrates that have complex surfaces or large surface areas.

Wen *et al.* [22] reported that a sol-gel method for HA and titania (TiO_2) coatings exhibited excellent bioactivity after immersion in simulated body fluid (SBF) that mimics human body fluid of a similar ion concentration and pH value to human blood plasma. In addition to enhancing titanium bioactivity, HA-titania coating is expected to increase the bonding strength and corrosion resistance. The surface morphology and microstructure of HA and titania coating before and after being immersed in SBF are presented in Figure 1 (a)-(d). It can be seen that the coating is dense, uniform and without cracks. Wen *et al.* also reported that after soaking in SBF, HA granules grow gradually.

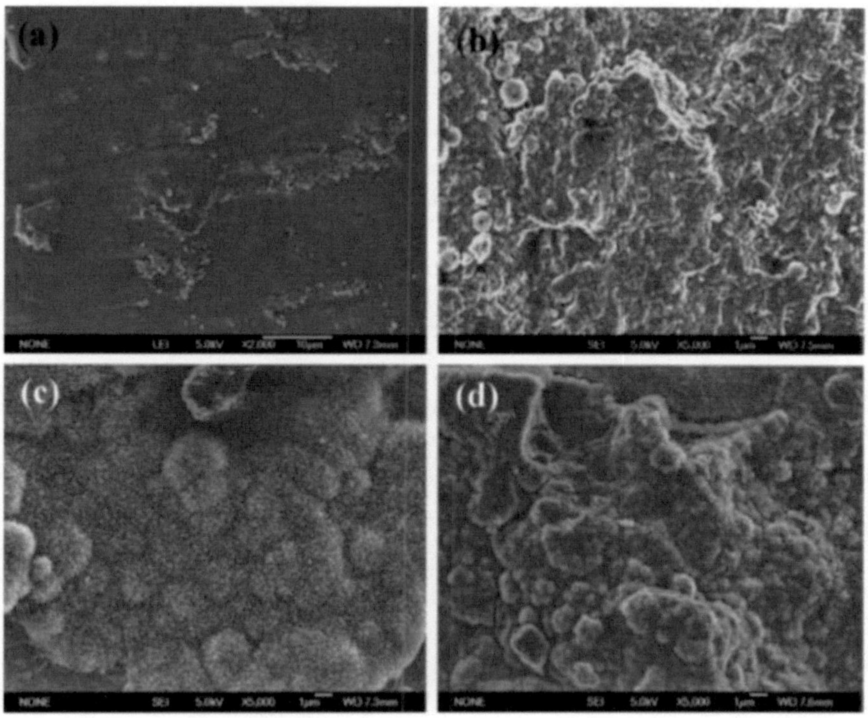

Figure 1. SEM micrographs of the surface morphology of HA/TiO_2 films after soaking in SBF for (a) 0 day, (b) 1 day, (c) 8 days, and (d) 15 days

5.2. Electrodeposition of materials

Electrodeposition is a coating method applied to the fabrication of computer chips and magnetic data storage. Recently, that has been rising interest in electrochemical deposition for tissue engineering applications due to its ability to coat complex 3D components.

Lopez-Heredia *et al.* [23] coated calcium phosphate onto porous titanium using the electrodeposition method. In the process, Ti, platinum mesh, and supersaturated calcium phosphate solution were used as the cathode, electrode and electrolyte, respectively. The ratio of Ca/P in the calcium phosphate coating was 1.65 and the coating thickness was 25 μm. The calcium phosphate coating was homogenous and covered the entire Ti surface. Moreover, they reported that the coating presented good adhesion to the underlying substrate. The electrodeposition of CaP showed that calcium phosphate enhances the adherence of cells.

Adamek *et al.* [24] succeeded in producing an HA coating on porous Ti6Al4V using electrodeposition. The intermediate layer between the bone and metallic implant was rough and porous. Large pores and nanolamellae were present within the HA layer. The flexibility of the electrodeposition technique for coating solid and porous metallic implants has acquired increasing interest due to this ability to enhance the bioactivity of bone implant materials.

5.3. Biomimetic creation of surfaces

There are two major steps involved in the biomimetic technique. The first step is to conduct a pre-treatment of the implant material surface to create a layer functional group that can induce formation of an effective apatite layer. Several studies have revealed that an apatite layer has not been formed on materials without any treatment prior to immersion [25]. Preliminary treatment includes, for example, hydrothermal, sol-gel, alkali heat treatment and micro-arc. The second step is to immerse the biomaterials into a simulated body fluid (SBF). In this step, the bone apatite layer is formed on the biomaterial's surface. The high apatite forming ability of titanium arises from the formation of a hydrated titanate surface layer during chemical treatment. The advantages of the biomimetic process include (i) flexibility in the control of the chemical composition and thickness of the coating, (ii) the formation of relatively homogenous bioactive bonelike apatite coatings, (iii) a lower processing temperature, and (iv) the ability to coat 3D geometries.

Wang *et al.* [25] used a modified biomimetic approach to improve the biocompatibility of porous titanium alloy scaffolds. In their experiment, porous Ti10Nb10Zr underwent an alkali heat treatment prior to soaking in SBF. Two NaOH concentrations of 5 M and 0.5 M were used, and the samples were soaked for 1 week. The surface morphologies of porous TiNbZr after alkali soaking and heat treatment revealed a nanofiber layer, that consisted of sodium titanate. Parameters that influenced the morphology and thickness of the sodium titanate were reaction temperature and NaOH concentration.

Calcium phosphate was successfully deposited on the surface of the porous TiNbZr. The calcium phosphate layer was uniform and homogenously spread onto the surface. Another biomimetic study, conducted by Habibovic *et al.* [26], indicated that a thick and homogenenous crystalline hyroxyapatite coating was deposited on all pores and resembled bone minerals.

An evaporation-based biomimetic coating was introduced by Duan *et al.* [27]. In their study, a supersaturated calcium phosphate was prepared by mixing NaCl, $CaCl_2$, HCl, $NH_4H_2PO_4$, tri(hydroxymethyl)aminomethane (Tris), and distilled water, which they termed the accelerated calcification solution (ACS). Calcium phosphate crystallites formed on the surface on dipping the samples into the ACS. The main component in the coating was octa-calcium-phosphate (OCP) and apatite was observed after soaking in SBF. The advantages of this method include (i) no surface etching is required, (ii) high supersaturations of the coating chemistry can be achieved, and (iii) tight control of the solutions is achieved [27].

5.4. Thermal spray

The thermal spray technique is a well-established and versatile technique that can be applied for a wide variety of coating materials, *i.e.*, metallic, non-metallic, ceramic, and polymeric. Thermal spray coated medical implants, such as orthopaedic and dental prostheses, have been commercially used. Thermal spray offers several advantages, such as the ability to coat low and high melting materials, a high deposition rate, and flexibility in coating 3D shape components. It is also cost effective [28]. Despite these advantages, some problems have been revealed after long term implantations using thermal spray coatings, such as delamination, resorption, biodegradation of the thick coating and mechanical instability [29]. Thus, improving the adhesion strength of thermal sprayed coatings is a major concern for bone or dental applications.

There are several types of thermal spraying; for example, plasma spray, flame spray, and cold spray [29]. Plasma spray is commonly applied to produce thick coatings for metallic corrosion protection. It is also flexible, due to its ability to coat different substrates. During plasma spraying, the precursor is atomised and injected into plasma jet, then accelerated towards the substrate with the aid of an inert carrier gas [30]. There are many parameters that must be controlled to produce a high quality coating.

Flame spray uses a combustion flame to melt the solid precursor. There is, additionally, another type of flame spray termed as high velocity oxygen fuel (HVOF). This technology is favourable due to its high spray velocity and the formation of a strong bond coating [30-32].

The thermal spray technique has been widely employed for HA coatings. The surface morphology of HA coatings obtained with various parameters of stand-off distance (SOD) and power are presented in Figure 2 (a)-(d). Sun *et al.* [28] reported that when the spray power increased, the crystallinity of HA decreased and the amorphous phase became more obvious. The effect of SOD indicated an inverse correlation with deposition efficiency. Several parameters that influence the deposition of HA are SOD, spray power, feedstock particle size and velocity.

Cold spray is a new member of the thermal spray family. This technique uses small particles of 1-50 μm. A supersonic jet of compressed gas is used to accelerate the particles. The advantage of using this technique is the ability to produce dense coatings and maintain the material chemistry and phase composition of the feedstock. Noppakun *et al.* [33] have applied cold spray technique to deposit HA-Ag/poly-esther-ether-ketone on glass slides. This study reported that cold spray was able to retain and elicit a coating functionality that was the same as the starting materials.

5.5. Physical vapor deposition

Physical vapor deposition (PVD) is a deposition method where materials are evaporated or sputtered, transferred and deposited onto the substrate surface. This physical process includes thermal evaporation or plasma-induced ion bombardment onto the sputtering target. A condensation or reaction of the coating materials then takes place on the substrate surface to form coatings. Variants of the PVD process include evaporation, ion plating, pulsed laser deposition and sputtering. The beneficial features of PVD are high coating density, high bio-adhesion strength, formation of multi-component layers, and low substrate temperature [34].

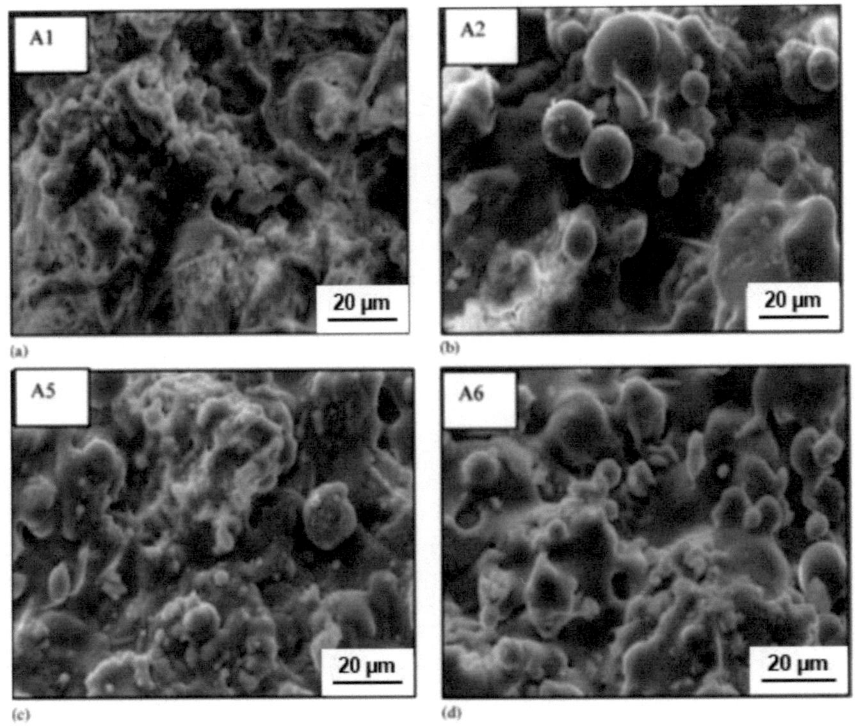

Figure 2. Surface morphology of HA coatings obtained by thermal spray method (a) 27.5 kW at 80 mm SOD, (b) 27.5 kW at 160 mm SOD, (c) 42 kW at 80 mm SOD, and (d) 42 kW at 160 mm SOD

Evaporation involves the thermal phase change from solid to vapor under vacuum conditions, in which evaporated atoms of a solid precursor placed in an open crucible can travel directly and condense onto the surface of a substrate [35]. A vacuum environment is used to minimize contamination [36]. Han *et al.* [37] have created an HA coating using electron beam evaporation and then incorporated silver by immersion into $AgNO_3$ solution. One advantage of this method is an improved bond strength between the coating and substrate. The ratio of Ca/P in the HA coating was 1.62 with a bond strength of 64.8 MPa, which was significantly higher than a plasma sprayed bond strength of 5.3 MPa [37].

Sputtering involves a process of ejecting neutral atoms from a target surface using energetic particle bombardment. The energetic particles used in the sputtering process are argon ions, which can be easily accelerated towards the cathode by means of an applied electric potential, hence bombarding the target, and ejecting neutral atoms from the target. These ejected atoms are then transferred and condense to the substrate to form a coating. Sputtering has been used in many applications such as the semiconductor, photovoltaic and automotive sectors. There are several sputtering methods, such as DC glow discharge, radio frequency (RF), ion beam sputtering (IBS), and reactive sputtering [36].

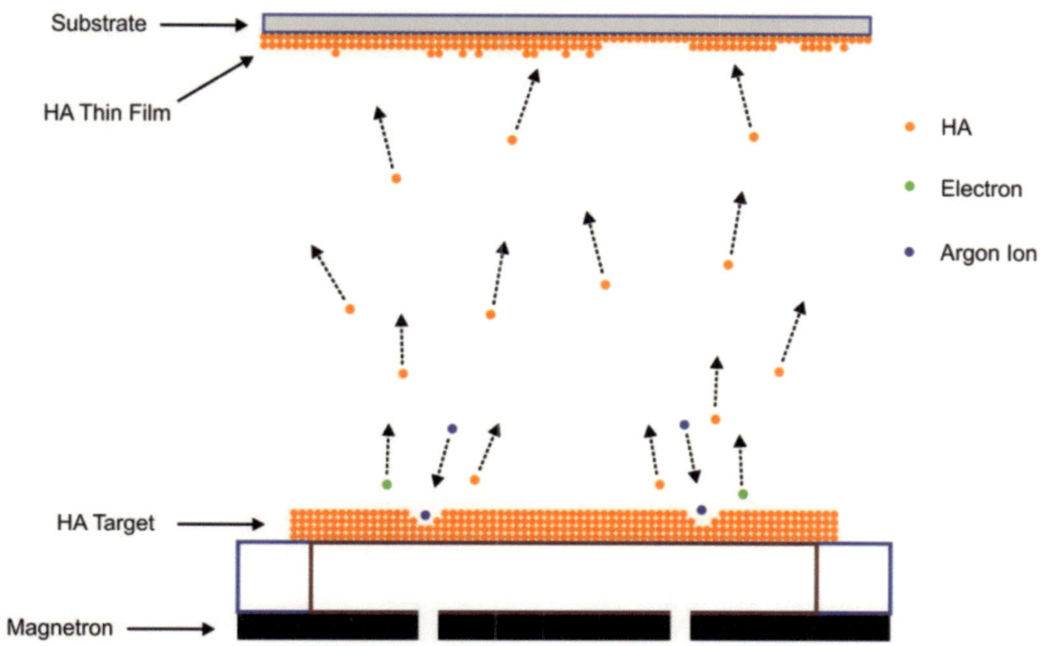

Figure 3. Schematic diagram of the sputtering mechanism

The simplest model for sputtering is the diode plasma, which consists of a pair of planar electrodes, an anode and a cathode, inside a vacuum system [37]. The sputtering target is mounted on the cathode. Application of the appropriate potential difference between the cathode and anode will ionize argon gas and create a plasma discharge. The argon ions will then be attracted and accelerated toward the sputtering target. Such ion bombardment on the target will displace some of the target atoms. This results in electron emission that will subsequently collide with gas atoms to form more ions that sustain the discharge [37]. Ion beam sputtering has disadvantages, such as a high capital investment cost (approximately one million dollars per machine), low deposition rates and a relatively small capacity per chamber batch [38]. Another type of sputtering employs radio frequency (RF) diodes that operate at high frequency.

Magnetron sputtering is one option to overcome the problems such as delamination and low bond strength that may arise with plasma spray methods. Magnetron sputtering enables lower pressures to be used, because a magnetic field allows trapping of the secondary electrons near the target. This induces more collisions with neutral gases and increases plasma ionisation. Figure 3 is a diagram of the magnetron sputtering mechanism. RF magnetron sputtering is an improved ion-sputtering method. It has also been noted that sputtered films possess higher adhesion to the substrate compared to the evaporation method.

A summary of the characteristics of the various coating techniques for calcium phosphate is presented in Table 2. Each technique has its own benefits and drawbacks. However, sputtering is a promising method due to its ability to produce dense and thin coatings, as well as provide good bond strength [39-41].

Techniques	Advantages	Disadvantages	Coating thickness
Sol-gel	Flexible in coating complex shapes,	Sometimes expensive	< 1 μm
Electrodeposition	Flexible in coating complex shapes. Low energy process, can be scaled down to deposition of a few atoms or scaled up to large dimensions.	Tends to crack	25 μm
Plasma Spray	Able to coat high and low melting materials. High deposition rate	Delamination and resorption. High temperature leads to decomposition	50 - 100 μm
Biomimetic	Flexible in coating complex shapes and flexible in controlling chemical composition of the coating. Homogenous.	The use of alkali heat treatment could reduce mechanical strength. Requires much time	10 - 30 μm
Sputtering	Dense, homogenous coating. Excellent adhesion	Needs annealing for crystalline structure	< 1 μm

Table 2. Summary of various techniques for calcium phosphate coatings

5.5.1. Properties of sputtered hydroxyapatite coatings

Coating thickness. The HA coating thickness varies. Molagic [42] succeeded in producing HA/ZrO_2 coatings with an average thickness of 3.2 μm. Hong *et al.* [43] manufactured a 500 nm thick coating of crystalline HA using magnetron sputtering. Ding [44] sputter deposited HA/Ti coatings with a film thickness of 3-7 μm onto a titanium substrate. Thian *et al.* [45] succeeded in incorporating silicon in hydroxyapatite (Si-HA) using magnetron sputtering and discovered its potential use as a bio-coating. The Si-HA film thickness was up to 700 nm.

Bond strength. An *in vitro* and *in vivo* experiment on coatings using the sputtering technique revealed coating detachment problems. Cooley *et al.* [46] reported that HA coatings were removed after 3 weeks of implantation. A bond layer coating was suggested to overcome this weak adhesion at the interface and subsequent delamination. Ievlev *et al.* [47] measured the adhesion strength of HA coatings with a sublayer and revealed that the adhesion strength was higher than coatings without a sublayer. Nieh *et al.* [48] used titanium as a pre-coat on Ti6Al4V and found strong bonding between the Ti layer and the HA coating.

Layered materials have previously been demonstrated to improve bonding between dissimilar materials. According to Ding [45], the top layer provides an excellent interaction with the surrounding tissue and promotes bone healing. A functionally graded coating (FGC) is an alternative method to enhance coating adhesion strength. Ozeki *et al.* [49] prepared an FGC of HA/Ti onto a metallic substrate. The coating thickness was 1 μm and consisted of 5 layers. The configuration of FGC was designed so that the HA was more dense near the surface, whilst the Ti was more dense near the substrate. The bonding strength using the FGC configuration was higher than using only HA, *i.e.*, 15.2 MPa and 8 MPa for the FGC and pure HA, respectively.

Elastic properties. Snyders *et al.* [50] manufactured HA via RF sputtering and revealed that the chemical composition influenced the elastic properties. As the Ca/P ratio decreased, the elastic modulus also decreased due to the insertion of Ca vacancies in the HA lattices.

5.5.2. Biological performance of sputtered hydroxyapatite coatings

The biological behaviour of biomaterials has been a fundamental criterion for successful candidate implant materials, along with their mechanical properties. The surface properties of a biomaterial play a significant role in the cell response. Thus, surface modification is an established strategy that has been used for biomedical applications due to its ability to enhance bioactivity. High cell density enhances bone formation. The cell adhesion behaviour and proliferation are influenced by several factors, such as pore size, porosity, and surface composition [51].

Thian *et al.* [45] carried out an *in vitro* test using a human osteoblast (HOB) cell model for a silicon incorporated hydroxyapatite (Si-HA) coating on titanium. The sample demonstrated an increase in metabolic activity compared to mono-HA coatings. Sputtered HA and Si coatings exhibited good differentiation of osteogenic cells and good biocompatibility. It was noted that the biological response was influenced by the crystallinity of the HA coatings. Sputtered composite coatings of HA with other compounds may provide additional advantages for implant performance. For instance, Chen *et al.* [52] incorporated silver into HA, conducted a cytotoxic and antibacterial test, and reported that the silver had an antibacterial effect since the bacterial attachment was reduced compared to coatings that did not contain silver.

6. Experimental methods

6.1. Design and preparation of titanium alloys

Tin and niobium were chosen as alloying elements because both metals are biocompatible and non-cytotoxic. The titanium alloy composition was designed using the molecular orbital DV-Xα method [53]. The calculation of the nominal composition of the alloys was based on two parameters, known as the bond order (Bo) and d-orbital energy level (Md). The parameter Bo is the covalent bond strength between titanium and an alloying element, while the parameter Md represents the d-orbital energy level of a transition alloying metal that correlates with the electro-negativity and the atomic radius of element. The list of Md and Bo values for each alloying elements (Ti, Nb and Sn) was obtained from a study conducted by Abdel Hady *et al.* [54].

Titanium alloys were fabricated using the powder metallurgy technique. Titanium powders (purity 99.7%), tin powders (purity 99.0%) and niobium powders (purity 99.8%) with particle sizes less than 45 μm were used. Each component was first weighted to give the desired composition of Ti14Nb4Sn. Ammonium hydrogen carbonate (NH_4HCO_3) was used as a space holder material. The particle size chosen was 300-500 μm in diameter.

The desired porosity and pore size were controlled by adjusting the initial weight ratio of NH_4HCO_3 to metal powders and the particle size of NH_4HCO_3. These components were mixed and blended in a planetary ball milling for 4 h with a weight ratio of ball to powder of 1:2 and a rotation rate of 100 rpm. A small amount of ethanol was employed during the mixing of the ammonium hydrogen carbonate with elemental metal powders to prevent segregation. After mixing the ammonium hydrogen carbonate with the metal powders, the mixture was pressed into green compacts in a 50 ton hydraulic press.

The green compacts were sintered at a pressure of 1.3 x 10^{-3} Pa using a vacuum furnace. Two steps of heat treatment were employed to produce porous structures. The first step was to burn out the space holder particles at 200°C for 2 h. The second step was to sinter the compacts at 1200°C for 10 h. Dense samples were prepared using powder metallurgy with the absence of space holder particles, and heat treatment was carried out at 1200°C. The dimensions of dense and porous titanium alloy samples were 9 mm in diameter and 2 mm in thickness for subsequent sample characterization. The sintering process was conducted at 1200°C for 10 h. A schematic diagram of the fabrication sequence for titanium alloys is presented in Figure 4.

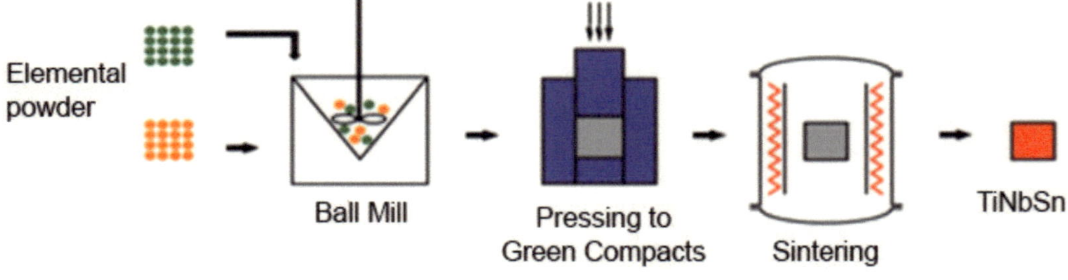

Figure 4. Schematic of Ti14Nb4Sn fabrication process by powder metallurgy route

Titanium alloy discs with 6 mm in diameter and 2 mm in thickness were gently wet grounded using (i) silicon carbide paper of 600 grit, (ii) followed by 1200 grit, and (iii) fine polished using 15, 9, 6, and 1 μm diamond compounds progressively. All metallic discs were then ultrasonically cleaned using ethanol for 5 min.

6.2. E-beam evaporation and sputtering

Silica thin films and nanocrystalline hydroxyapatite coatings were successively deposited onto the prepared titanium alloy substrates by e-beam evaporation and sputtering techniques. A HV thin film deposition system (CMS-18 Kurt J. Lesker, USA) was used. Both the e-beam evaporation and the sputtering processes were performed at room temperature. The base pressure of the system was 6.6 x 10^{-6} Pa.

A 200 nm SiO_2 thin film was deposited at a working pressure of 6.6 x 10^{-4} Pa and a deposition rate of 10 nm/s. During the sputtering process, the working pressure was set at 0.8 Pa. The sputtering power was 90 W. The distance between the substrate and sputtering target was kept at 30 cm. During deposition, the substrate holder rotated in order to achieve uniform coating. Heat treatment of samples was performed at 500°C for 2 h in a vacuum furnace.

6.3. Characterization

The elemental composition was analyzed using an energy dispersive X-ray spectrometer (EDS, Oxford instruments INCA suite v.4.13) interfaced with a field-emission scanning electron microscope (FE-SEM, ZEISS SUPRA 40 VP) operated at 15 kV. Surface morphology of the samples was observed using scanning electron microscopy, and phase identification was performed using the X-ray diffraction method (XRD, Bruker D8 Advance), operated with CuK_α radiation in the Bragg-Brentano mode at a scanning rate of 0.5°/min over a 2θ range of 30-80°. Phase analysis was conducted using the database PDF-2 version 2005.

The porosity of the scaffold was characterised by gravimetry using the formula [13]:

$$\varepsilon = \left(1 - \frac{\rho}{\rho_s}\right) \times 100 \qquad (1)$$

where ρ and ρ_s are the actual and theoretical densities of the porous alloy, respectively.

7. Results and discussion

7.1. Physico-chemical properties of the Ti14Nb4Sn alloy

The X-ray diffraction pattern of sintered Ti14Nb4Sn is shown in Figure 5. Alpha peaks were observed at 39.0° and 40.5°, which are indexed as the reflection planes (101) and (103), while β peaks were observed at 38.5°, which is indexed as (110). The titanium alloy consisted of both α and β phases. Weak niobium peaks were also detected, while tin was not detected. Elemental analysis using EDS was performed concurrently with the SEM examination to identify the chemical composition of the samples. The EDS analyses verified that the alloy composition corresponded to Ti14Nb4Sn.

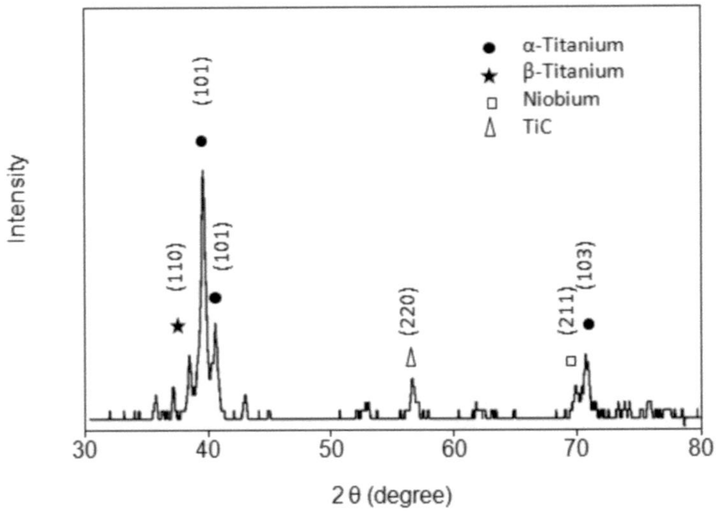

Figure 5. XRD pattern of sintered Ti14Nb4Sn alloy

SEM images of fabricated porous titanium alloys showed a combination of both macropores and micropores on the surface, as shown in Figure 6 (a)-(f). The micropore size ranged from 0.5 to 10 μm, while the macropore size ranged from 50 to 700 μm. Samples with greater porosity exhibited more interconnected features, more accessible inner surfaces, and interpenetrated macropores. It is believed that the optimal pore size to ensure vascularization and bone in-growth is 50-400 μm [13]. Compared to other studies, fabrication of Ti10Nb10Zr alloy resulted in pore sizes ranging from 300 to 800 μm since the size of the space-holder particles was set to be 500-800 μm [13].

Usually there are two types of pores when using the space-holder method to fabricate titanium alloys: (i) macro-pores determined by the size of the space holder particles, and (ii) micro-pores determined by the dimension of the titanium powder particles. The micropores can be designed to allow the scaffold to be impregnated with functional coatings or therapeutic agents.

Porosity enhances the interlocking processes for the stability and immobility of the new implant, often referred to as stabilization and fixation of the implant. The porosity is influenced by several factors, namely the particle size of the metallic powder and the sintering pressure [13]. The porosity of the samples ranges from 55 to 80%. The optimum porosity of the implant for bone in-growth is in the range of 50-90%. It has been noted that the porosity level of an implant should be selected to provide the optimum mechanical behaviour, since porosity has a dominant and adverse influence on the strength of a porous material.

The pore connectivity, which can be determined by percolation theory, is a crucial parameter that determines successful bone in-growth. Connectivity between the pore provides sufficient area for physiological fluid to flow throughout the new tissue that enhances nutrient transportation. The images in Figure 6 exhibit variation in pores connectivity. High porosity results in high pore interconnectivity, *i.e.*, samples with 80% porosity exhibit high pore connectivity.

Figure 6. Morphology of porous Ti14Nb4Sn alloys with different porosity: (a) 55%, (b) 60%, (c) 70%, (d) 72%, (e) 75%, and (f) 80%

7.2. Physico-chemical properties of sputtered hydroxyapatite coated titanium alloys

The application of SiO_2 as a bond layer between the substrate and the coating should improve coating adhesion to the substrate. One advantage of using silica is its influence on the bone mineralization process. Li *et al.* applied silica onto a titanium surface using a sol gel process and demonstrated its bioactivity [11]. Hong *et al.* [4] conducted an *in vitro* bioactivity test on bioactive ceramic glass with higher silicon content and revealed a superior mineralization capability. Thian *et al.* [46] succeeded in incorporating silicon into hydroxyapatite (Si-HA) using magnetron sputtering, and reported that higher silica content was beneficial for biomedical applications due to its higher corrosion resistance.

Figure 7 shows the surface topography of the HA-silica coating on titanium alloy Ti14Nb4Sn. The 2 μm thick hydroxyapatite coating and 200 nm thick SiO_2 film were deposited onto the titanium alloy using RF magnetron sputtering and e-beam evaporation, respectively. The HA coating was homogenous, which is characteristic of thin films deposited by sputtering. However, some cracks on the surface were observed. Some morphological features of rough coatings with some cracks could be advantageous for bone implant applications since this morphology could act as an anchorage for tissue growth.

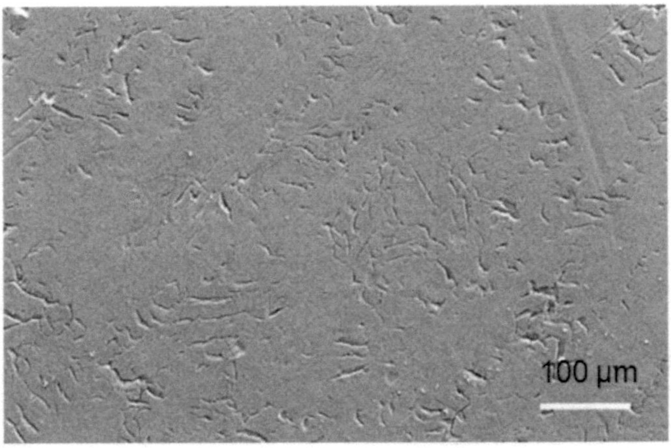

Figure 7. Morphology of HA-SiO_2 coated Ti14Nb4Sn alloy

The XRD pattern of the HA-SiO_2 coated titanium alloys is shown in Figure 8. The identified phases were hydroxyapatite, CaO.SiO_2.TiO_2, calcium pyrophosphate, $CaTiO_3$ and titanium. After annealing, the crystalline phase of HA was present at $2\theta = 30°$ which matches the (107) plane. A peak corresponding to CaO.SiO_2.TiO_2 phase was also observed at 43.5° and indexed as (223). In addition, the peak confirms the presence of the silica phase. The phase CaO, *i.e.*, in CaO.SiO_2.TiO_2 observed in the XRD pattern could be related to the partial decomposition of hydroxyapatite during the deposition process.

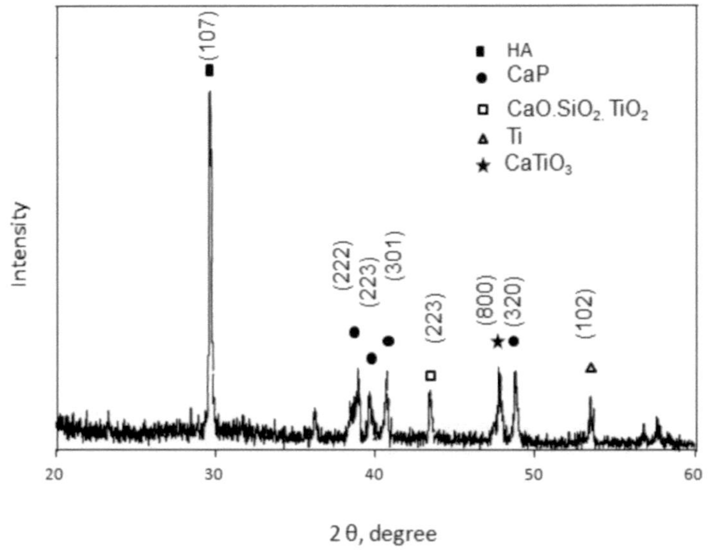

Figure 8. XRD patterns of HA-SiO_2 coatings on Ti14Nb4Sn alloy

The titanium alloys are likely to be oxidized during the annealing process. Therefore, TiO_2 appeared in the $CaO.SiO_2.TiO_2$ phase and $CaTiO_3$ phase. The $CaTiO_3$ peak was detected at 47.8° with an orientation of (800). The four peaks at 38.8°, 39.7°, 40.5° and 48.9° corresponding to calcium phosphate ($Ca_2P_2O_7$) are indexed as the reflection planes (222), (223), (301) and (320), respectively. However this phase might have higher solubility compared to HA. It is possible that during the sputtering process not all components of the HA target were sputtered and transferred onto the substrate. The titanium peak was present at 53.5° and indexed as (102). The results indicated that HA coatings using magnetron sputtering could produce the crystalline apatite phase.

8. Conclusions

This chapter describes the importance of developing a bioactive titanium alloy scaffold for bone tissue engineering applications. Ti14Nb4Sn alloy was designed and then fabricated using powder metallurgy method. The porosity ranged from 55 to 80% with pore sizes of 100-600 μm.

Powder metallurgy that employed the space-holder sintering method was successful in fabricating samples for biomedical implant studies. The method produced porous structures that (i) enable better fixation, (ii) lower elastic modulus to match the properties of natural bone, and (iii) construct morphologies that mimic the features of natural bone structures.

To further enhance the biocompatibility of titanium alloys, 2 μm thick hydroxyapatite and 200 nm thick SiO_2 coatings were deposited onto Ti alloys using e-beam evaporation and RF magnetron sputtering. SEM images showed that the microstructure of the hydroxyapatite coating is homogenous, with some cracks appearing on its surface. XRD results confirmed that the coatings consisted of an HA phase with some $CaO.SiO_2.TiO_2$, $CaTiO_3$ and phases. Silica was also present in the XRD spectrum, which corresponds to the $CaO.SiO_2.TiO_2$ phase. It was demonstrated that the e-beam evaporation and magnetron sputtering methods are suitable for depositing silica and hydroxyapatite coatings. The hydroxyapatite-silica configuration may be useful for biomedical implants, as it provides better adhesion strength for rapid osseointegration acceleration. Further study will focus on the biological response of these coatings.

Acknowledgements

CW acknowledges the financial support from the Australian Research Council (ARC) through the ARC Discovery Project DP110101974.

Author details

Kun Mediaswanti[1], Cuie Wen[1], Elena P. Ivanova[2], Christopher C. Berndt[1,3] and James Wang[1*]

*Address all correspondence to: jawang@swin.edu.au

1 Industrial Research Institute Swinburne, Faculty of Engineering and Industrial Sciences, Swinburne University of Technology, Hawthorn, Australia

2 Faculty of Life and Social Sciences, Swinburne University of Technology, Hawthorn, Australia

3 Adjunct Professor, Materials Science and Engineering, Stony Brook University, Stony Brook, New York, USA

References

[1] Jenkins GW, Kemnitz CP, Tortora GJ. Anatomy and Physiology from Science to Life. New Jersey: John Wiley & Sons, Inc.; 2007.

[2] Currey JD. Bones: Structure and Mechanics. New Jersey: Princeton University Press; 2002.

[3] Bauer T, Muschler G. Bone graft materials. An overview of the basic science. Clinical Orthopaedic Related Research 2000;371:10-27.

[4] Brånemark R, Brånemark PI, Rydevik B, Myers RR. Osseointegration in skeletal reconstruction and rehabilitation: a review. Journal of Rehabilitation Research and Development. 2001;38(2):175-81.

[5] Jafari SM, Bender B, Coyle C, Parvizi J, Sharkey PF, Hozack WJ. Do tantalum and titanium cups show similar results in revision hip arthroplasty? Clinical Orthopaedics and Related Research. 2010;468(2):459-65.

[6] Long M, Rack HJ. Titanium alloys in total joint replacement - A materials science perspective. Biomaterials. 1998;19(18):1621-39.

[7] Kujala S, Ryhanen J, Danilov A, Tuukkanen J. Effect of porosity on the osteointegration and bone ingrowth of a weight-bearing nickel–titanium bone graft substitute. Biomaterials. 2003;24:4691–7.

[8] LeGeros RZ, Craig RG. Strategies to affect bone remodeling: Osteointegration. Journal of Bone and Mineral Research. 1993;8(Suppl. 2):S583-S96.

[9] Hollister SJ. Porous scaffold design for tissue engineering. Nature Materials. 2005;4(7):518-24.

[10] Li Y, Wong C, Xiong J, Hodgson P, Wen C. Cytotoxicity of titanium and titanium alloying elements. Journal of Dental Research. 2010;89(5):493-7.

[11] Okazaki Y, Rao S, Asao S, Tateishi T, Katsuda SI, Furuki Y. Effects of Ti, Al and V concentrations on cell viability. Materials Transactions, JIM. 1998;39(10):1053-62.

[12] Obbard EG, Hao YL, Akahori T, Talling RJ, Niinomi M, Dye D, et al. Mechanics of superelasticity in Ti-30Nb-(8-10)Ta-5Zr alloy. Acta Materialia. 2010;58(10):3557-67.

[13] Wang XJ, Li YC, Xiong JY, Hodgson PD, Wen CE. Porous TiNbZr alloy scaffolds for biomedical applications. Acta Biomaterialia. 2009;5(9):3616-24.

[14] Xiong JY, Li YC, Hodgson PD, Wen CE. Mechanical properties of porous Ti-26Nb alloy for regenerative medicine. Fan JH, Chen HB. Advances in Heterogeneous Material Mechanics. 2008: 630-634.

[15] Chen X, Nouri A, Li Y, Lin J, Hodgson PD, Wen C. Effect of surface roughness of Ti, Zr, and TiZr on apatite precipitation from simulated body fluid. Biotechnology and Bioengineering. 2008;101(2):378-87.

[16] Kokubo T. Bioactive glass ceramics: properties and applications. Biomaterials. 1991;12(2):155-63.

[17] Li J, Habibovic P, Yuan H, van den Doel M, Wilson CE, de Wijn JR, et al. Biological performance in goats of a porous titanium alloy-biphasic calcium phosphate composite. Biomaterials. 2007;28(29):4209-18.

[18] Suzuki Y, Nomura N, Hanada S, Kamakura S, Anada T, Fuji T, et al. Osteoconductivity of porous titanium having young's modulus similar to bone and surface modification by OCP. Key Engineering Materials.2007;330-332(II): 951-4.

[19] Hu S, Li S, Yan Y, Wang Y, Cao X. Apoptosis of cancer cells induced by HAP nanoparticles. Journal Wuhan University of Technology, Materials Science Edition. 2005;20(4):13-5.

[20] Hu Q, Tan Z, Liu Y, Tao J, Cai Y, Zhang M, et al. Effect of crystallinity of calcium phosphate nanoparticles on adhesion, proliferation, and differentiation of bone marrow mesenchymal stem cells. Journal of Materials Chemistry. 2007;17(44):4690-8.

[21] Dumbleton J, Manley MT. Hydroxyapatite-coated prostheses in total hip and knee arthroplasty. Journal of Bone and Joint Surgery - Series A. 2004;86(11):2526-40.

[22] Wen CE, Xu W, Hu WY, Hodgson PD. Hydroxyapatite/titania sol-gel coatings on titanium-zirconium alloy for biomedical applications. Acta Biomaterialia. 2007;3(3): 403-10.

[23] Lopez-Heredia MA, Sohier J, Gaillard C, Quillard S, Dorget M, Layrolle P. Rapid prototyped porous titanium coated with calcium phosphate as a scaffold for bone tissue engineering. Biomaterials. 2008;29(17):2608-15.

[24] Adamek G, Jakubowicz J. Mechanoelectrochemical synthesis and properties of porous nano-Ti-6Al-4V alloy with hydroxyapatite layer for biomedical applications. Electrochemistry Communications. 2010;12(5):653-6.

[25] Wang X, Li Y, Hodgson PD, Wen C. Biomimetic modification of porous TiNbZr alloy scaffold for bone tissue engineering. Tissue Engineering - Part A. 2010;16(1):309-16.

[26] Habibovic P, Barrère F, Van Blitterswijk CA, De Groot K, Layrolle P. Biomimetic hydroxyapatite coating on metal implants. Journal of the American Ceramic Society. 2002;85(3):517-22.

[27] Duan K, Tang A, Wang RZ. A new evaporation-based method for the preparation of biomimetic calcium phosphate coatings on metals. Materials Science & Engineering C-Biomimetic and Supramolecular Systems. 2009;29(4):1334-7.

[28] Sun L, Berndt CC, Grey CP. Phase, structural and microstructural investigations of plasma sprayed hydroxyapatite coatings. Materials Science and Engineering A. 2003;360(1-2):70-84.

[29] Kweh SWK, Khor KA, Cheang P. An in vitro investigation of plasma sprayed hydroxyapatite (HA) coatings produced with flame-spheroidized feedstock. Biomaterials. 2002;23(3):775-85.

[30] Saber-Samandari S, Berndt CC. IFTHSE Global 21: Heat treatment and surface engineering in the twenty-first century: Part 10 - Thermal spray coatings: A technology review. International Heat Treatment and Surface Engineering. 2010;4(1):7-13.

[31] Tang Q, Brooks R, Rushton N, Best S. Production and characterization of HA and Si-HA coatings. Journal of Materials Science-Materials in Medicine. 2010;21(1):173-81.

[32] Richard C, Kowandy C, Landoulsi J, Geetha M, Ramasawmy H. Corrosion and wear behavior of thermally sprayed nano ceramic coatings on commercially pure Titanium and Ti-13Nb-13Zr substrates. International Journal of Refractory Metals and Hard Materials. 2010;28(1):115-23.

[33] Sanpo N, Tan ML, Cheang P, Khor KA. Antibacterial property of cold-sprayed HA-Ag/PEEK coating. Journal of Thermal Spray Technology. 2009;18(1):10-5.

[34] Goswami R, Jana T, Ray S. Transparent polymer and diamond-like hydrogenated amorphous carbon thin films by PECVD technique. Journal of Physics D: Applied Physics. 2008;41(15).

[35] Rossnagel S. Sputtering and Sputter Deposition. In: Seshan K, (ed.). Handbook of Thin-Film Deposition Processes and Techniques - Principles, Methods, Equipment and Applications. California: William Andrew Publishing; 2002.

[36] Liu XY, Chu PK, Ding CX. Surface modification of titanium, titanium alloys, and related materials for biomedical applications. Materials Science & Engineering R-Reports. 2004;47(3-4):49-121.

[37] Ivanova E, Wang J, Truong V, Kemp A, Berndt C, Crawford R. Bacterial Attachment onto the Surfaces of Sputter-prepared Titanium and Titanium-based Nanocoatings. In: Zacharie B. and Jerome T, (ed.) Advances in Nanotechnology: Nova Science Publisher; 2011.

[38] MacMillan I. Advances in sputtering benefit coating costs. Laser Focus World. 2009;45(4):41-4.

[39] Brohede U, Zhao S, Lindberg F, Mihranyan A, Forsgren J, Strømme M, et al. A novel graded bioactive high adhesion implant coating. Applied Surface Science. 2009;255(17):7723-8.

[40] Wan T, Aoki H, Hikawa J, Lee JH. RF-magnetron sputtering technique for producing hydroxyapatite coating film on various substrates. Bio-Medical Materials and Engineering. 2007;17(5):291-7.

[41] Pichugin VF, Surmenev RA, Shesterikov EV, Ryabtseva MA, Eshenko EV, Tverdokhlebov SI, et al. The preparation of calcium phosphate coatings on titanium and nickel-titanium by rf-magnetron-sputtered deposition: Composition, structure and micromechanical properties. Surface and Coatings Technology. 2008;202(16):3913-20.

[42] Molagic A. Structural characterization of TiN/HAp and ZrO_2 /HAp thin films deposited onto Ti-6Al-4V alloy by magnetron sputtering. UPB Scientific Bulletin, Series B: Chemistry and Materials Science. 2010;72(1):187-94.

[43] Hong Z, Mello A, Yoshida T, Luan L, Stern PH, Rossi A, et al. Osteoblast proliferation on hydroxyapatite coated substrates prepared by right angle magnetron sputtering. Journal of Biomedical Materials Research - Part A. 2010;93(3):878-85.

[44] Ding SJ. Properties and immersion behavior of magnetron-sputtered multi-layered hydroxyapatite/titanium composite coatings. Biomaterials. 2003;24(23):4233-8.

[45] Thian ES, Huang J, Best SM, Barber ZH, Bonfield W. A new way of incorporating silicon in hydroxyapatite (Si-HA) as thin films. Journal of Materials Science-Materials in Medicine. 2005;16(5):411-5.

[46] Cooley DR, Vandellen AF, Burgess JO, Windeler AS. The advantages of coated titanium implants prepared by radiofrequency sputtering from hydroxyapatite. Journal of Prosthetic Dentistry. 1992;67(1):93-100.

[47] Ievlev VM, Domashevskaya EP, Putlyaev VI, Tret'yakov YD, Barinov SM, Belonogov EK, et al. Structure, elemental composition, and mechanical properties of films prepared by radio-frequency magnetron sputtering of hydroxyapatite. Glass Physics and Chemistry. 2008;34(5):608-16.

[48] Nieh TG, Jankowski AF, Koike J. Processing and characterization of hydroxyapatite coatings on titanium produced by magnetron sputtering. Journal of Materials Research. 2001;16(11):3238-45.

[49] Ozeki K, Yuhta T, Fukui Y, Aoki H, Nishimura I. A functionally graded titanium/hydroxyapatite film obtained by sputtering. Journal of Materials Science-Materials in Medicine. 2002;13(3):253-8.

[50] Snyders R, Bousser E, Music D, Jensen J, Hocquet S, Schneider JM. Influence of the chemical composition on the phase constitution and the elastic properties of RF-sputtered hydroxyapatite coatings. Plasma Processes and Polymers. 2008;5(2):168-74.

[51] Socol G, Macovei AM, Miroiu F, Stefan N, Duta L, Dorcioman G, et al. Hydroxyapatite thin films synthesized by pulsed laser deposition and magnetron sputtering on PMMA substrates for medical applications. Materials Science and Engineering B: Solid-State Materials for Advanced Technology. 2010;169(1-3):159-68.

[52] Chen W, Liu Y, Courtney HS, Bettenga M, Agrawal CM, Bumgardner JD, et al. In vitro anti-bacterial and biological properties of magnetron co-sputtered silver-containing hydroxyapatite coating. Biomaterials. 2006;27(32):5512-7.

[53] Morinaga M, Yukawa H. Alloy design with the aid of molecular orbital method. Bulletin of Materials Science. 1997;20(6):805-15.

[54] Abdel-Hady M, Hinoshita K, Morinaga M. General approach to phase stability and elastic properties of β-type Ti-alloys using electronic parameters. Scripta Materialia. 2006;55(5):477-80.

Chapter 3

Modelling of Microstructural Evolution of Titanium During Diffusive Saturation by Interstitial Elements

Yaroslav Matychak, Iryna Pohrelyuk,
Viktor Fedirko and Oleh Tkachuk

Additional information is available at the end of the chapter

http://dx.doi.org/10.5772/54626

1. Introduction

Titanium and titanium alloys are the most promising structural materials for the products of the contemporary aircraft and spacecraft engineering, medicine. The complex of characteristics of such products strongly depends on the properties of their surface layers. One of the efficient method of their hardening is the thermodiffusive saturation with interstitial elements, in particular nitrogen or oxygen (Fedirko & Pohrelyuk, 1995; Panasyuk, 2007). Such high-temperature interaction with these interstitial elements is accompanied by not only the formation and growth of a nitride or oxide film, but also the significant dissolution of nitrogen or oxygen in the base metal. The competition of these processes complicates significantly the study of the kinetics and mechanism of such an interaction. In this case, useful information can be obtained from results of an investigation of the high-temperature interaction of titanium in an atmosphere with a decreased nitrogen or oxygen pressure, which simultaneously generates practical interest, because deep diffusion layers without a continuous nitride or oxide film on a titanium surface can be formed (Fedirko & Pohrelyuk, 1995; Panasyuk, 2007). The incubation period of formation of such a film depends to a large degree on the partial gas pressure and saturation temperature. Attempts to choose purposefully an optimal nitrogen or oxygen pressure and temperature–time parameters of such a thermochemical treatment failed. This is due to the complexity and diversity of the interactions of titanium with rarefied gas-containing atmospheres, the absence of data on parameters that characterize surface phenomena, and a large spread (up to two orders of magnitude) of available data on the diffusion coefficient of nitrogen or oxygen in titanium (Panasyuk, 2007; Metin, 1989; Kofstad, 1966). This is why investigations (experimental and theoretical) aimed at elucidating the kinetic regularities and peculiarities of the distribution of interstitial elements in a surface layer, which determines the changes of its physicomechanical characteristics, are urgent.

Diffusive processes determine changes of properties of surface layers of the structural materials in many cases, for example, in the process of their thermochemical treatment or in the conditions of operation at high temperature. However, the diffusion in solids is often accompanied by the structural phase transformations. These processes are interconnected and interdependent: diffusion of the elements can stimulate structural phase transformations, and the latter change the conditions of diffusion. It is difficult to describe these processes analytically. Titanium, which undergoes the polymorphic transformation at $T_{\alpha\leftrightarrow\beta} = 882\ ^0C$ (Fromm & Gebhardt, 1976), is interesting for such theoretical and experimental investigations, in particular its high-temperature interaction with nitrogen or oxygen. Due to high affinity of these elements with titanium nitride or oxide layer forms and grows on the surface. Unlike many alloying elements, in particular vanadium, molybdenum, which are β-stabilizers, above mentioned interstitial elements are α-stabilizers, which can stimulate structural phase transformations in titanium. The microstructural evolution during $\alpha \leftrightarrow \beta$ phase transformation as a result of migration of β-stabilizers is presented in (Malinov et al., 2003). However, the authors did not take into consideration the role of nitrogen as α-stabilizer in the structural transformations. It was demonstrated (Matychak, 2009) in the studies of interconnection of nitrogen diffusion and structural phase transformations during high-temperature nitriding that, in particular, under the rarefied atmosphere, the continuous nitride layer on the surface was absent for a long time.

The aim of work is:

- to establish the kinetic peculiarities of interaction of titanium with the interstitial element A (nitrogen or oxygen) at the temperature lower and higher than temperature of allotropic transformation $T_{\alpha\leftrightarrow\beta}$;
- to investigate experimentally and model analytically the process of diffusive saturation of α-titanium with the interstitial element from a rarefied atmosphere taking into account the surface phenomena;
- to model the interdependence of the processes of external supply of the interstitial element to the surface and its chemosorption with diffusive dissolution and segregation on defects, caused by the chemical interaction with the titanium atoms;
- to estimate the influence of temperature-time parameters of treatment on the depth of diffusion zone and change of its microhardness;
- to establish the kinetic peculiarities of diffusive saturation of titanium with the interstitial element caused by the structural phase transformations.

2. Thermodiffusion saturation of titanium with interstitial elements from a rarefied atmosphere at $T<T_{\alpha \leftrightarrow \beta}$

2.1. Physicomathematical model

2.1.1. Phenomenology of surface phenomena

Let us consider the interaction of α-titanium with a rarefied gas atmosphere in a temperature range which is below the temperature of the $\alpha \leftrightarrow \beta$ allotropic transformation. In such a system, peculiarities of the interaction predominantly manifest themselves on the titanium surface as a result of adsorption, chemisorption, chemical reactions, generation of point defects, and the formation of two-dimensional structures. Along with phase formation, which includes these processes on the surface, the transfer of the interstitial element in the depth of titanium, i.e., its diffusive saturation, plays an important role. Experimental data indicate that, for rather long exposures, at certain rarefaction of the interstitial element, only islands of a nitride or oxide film, rather than a continuous film, are formed on the titanium surface (Fedirko & Pohrelyuk, 1995). In this case, the kinetics of saturation is sensitive to the interstitial element transfer to the surface of titanium and the intensity of surface processes. Thus, the surface interstitial element concentration depends on time. The defectiveness of the metal and its influences on the diffusion activity and reactivity of the interstitial element also play an important role. Due to lattice defects, in particular vacancies, dislocations of the surface layer, the probability of inequilibrium segregations of the interstitial element increases as a result of the chemical interaction with titanium, which introduces changes in the diffusive saturation of titanium with the interstitial element. That is why it is incorrect to describe analytically the kinetics of saturation with the known Fick's equation by setting constant values of the surface concentration (the first boundary-value task). This indicates the actuality and importance of an adequate choice of boundary conditions for the formulation of the corresponding diffusion problem. To do this, it is necessary to have a clear notion of the interrelation of the physico-chemical processes on a surface and near it.

The interaction of titanium with a rarefied gas atmosphere can be schematically illustrated by following processes with relevant parameters characterizing them (Fig. 1):

a. transport of the interstitial element molecules to a metal surface followed by their physical adsorption, dissociation, and chemisorption (the mass transfer coefficient h, cm/sec);

b. segregation of the interstitial element on defects in a contact layer (with a mass capacity ϖ, cm) as a result of the chemical interaction with the metal (the rate of reaction k, cm/sec);

c. diffusion of the interstitial element in α-titanium (diffusion coefficient D, cm^2/sec).

Processes enumerated in clause a) can be interpreted as a two-stage reaction which consists of a diffusive stage, described by the constant rate h_D, and a stage of chemisorption at a constant rate h_R. Then, according to the law of summation of kinetic resistances, we have

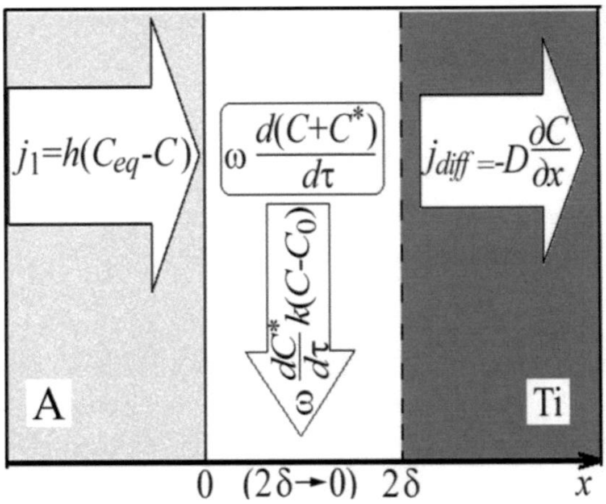

Figure 1. Scheme of mass fluxes in the Ti–A system

$h^{-1}=h_D^{-1}+h_R^{-1}$. The introduced kinetic parameters (P_i=D, h, k) of the model representation characterize the aforementioned thermoactivated physicochemical processes with the corresponding activation energies (E_i), according to the dependence $P_i=P_{0i}\exp(-E_i/RT)$. The effective parameters h and k depend not only on temperature, but also largely on the partial pressure of interstitial element and defectiveness of the material. That is why they are usually calculated from specific experimental data of the kinetics of saturation. In particular, the experimental data for the spatial distribution of the interstitial element in surface layer of titanium after various exposure time (τ_1 and τ_2) allow to use a graphical method to compute the mass transfer coefficient h and the surface content (C_{eq}) of nitrogen which is in equilibrium with the atmosphere (Fig. 2) (Matychak et al., 2009).

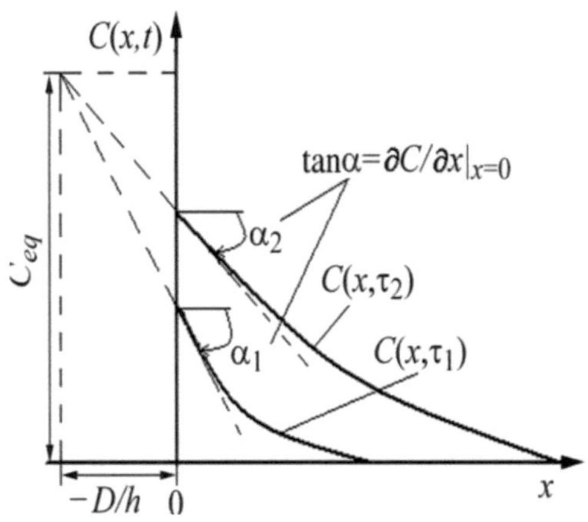

Figure 2. Graphical method for the determination of the h and C_{eq} parameters

Let us represent the inequilibrium processes mentioned in clauses (a) – (c) in the formulation of the diffusion problem through the adequate setting of the boundary conditions of mass exchange on the surface.

2.1.2. Mathematical description

Since the aim of the diffusive saturation of a titanium sample is primarily to harden its surface layer, as an object of the analytic investigation of the kinetics of this process, we chose a half-space ($0 \le x < \infty$) with the initial (τ=0) interstitial element concentration $C(x,\tau=0)=C_0$. For the calculation of the concentration of dissolved interstitial element in the titanium sample it is need to solve Fick's diffusion equation considering initial and boundary conditions (Matychak et al., 2007):

$$D\partial^2 C(x,\tau)/\partial x^2 = \partial C(x,\tau)/\partial\tau \quad for \quad \tau>0,\ 0<x<\infty,\ C(x,0)=C(\tau,\infty)=C_0, \qquad (1)$$

$$\omega \cdot dC/d\tau = h(C_{eq}-C) - k(C-C_0) + D\partial C/\partial x \qquad for \quad x=+0. \qquad (2)$$

Here C_{eq} is a quasiequilibrium surface concentration of the interstitial element, which depends on its partial pressure in the atmosphere.

The boundary condition (2) was proposed on the basis of notions of a contact layer with a thickness 2δ between the metal and the environment, in which processes of migration of an impurity and the chemical reaction of the first order (Fig. 1) occur (Prytula et al., 2005). Using a mathematical procedure (Fedirko et al., 2005; Matychak, 1999), this layer was replaced by an imaginary layer of zero thickness (2δ→ 0) with a mass capacity ω. For such a transition, we introduced averaged characteristics of the contact layer, specifically the surface concentration of the impurity $C(+0,\tau)$. Note that neglecting the contact layer ω=0, and, correspondingly, k = 0, from Eq.(2) we obtain the typical boundary condition of mass exchange of the third kind (Raichenko, 1981):

$$-D\partial C/\partial x\big|_{x=0} = h[C_{eq} - C(0,\tau)] \qquad (3)$$

or even the simpler condition (D/h →0) of the first kind:

$$C(0,\tau) = \lim_{x\to+0} C(x,\tau) = C_{eq} = const \qquad (4)$$

Let us point at the characteristic peculiarities of the proposed generalized boundary condition (2), which distinguishes it from the quasistationary boundary condition (3). The latter one reflects to a certain extent the real situation of the asymptotic approximation of the surface concentration to its equilibrium value. At the same time, according to condition (3), all atoms adsorbed on the surface diffuse into the metal and are distributed in compliance with the law of diffusion. That is why, according to condition (3), when $D\to 0$, we have $C(0,\tau)=C_{eq}$. That is, the surface concentration becomes equilibrium instantaneously and is independent of time. Thus, from the proposed generalized nonstationary boundary condition (2), in the absence of diffusion $D\to 0$ of the impurity in the volume of the metal, we have the following time dependence of its surface concentration:

$$C(0,\tau)=\frac{hC_{eq}+kC_0}{h+k}-\frac{h\left(C_{eq}-C_0\right)}{h+k}\exp\left[-\frac{(h+k)\tau}{\omega}\right], \tag{5}$$

which is determined by the intensity of surface processes. One more peculiarity of proposed non-stationary condition (2) concerns the action of the operator $d/d\tau$, which describes the kinetics of accumulation of the interstitial element in the vicinity of the interface. In particular, the difference between the flux j_1 of the interstitial element from the environment to the surface ($x=-0$) and its diffusion flux $j=j_{diff}$ in the metal ($x=+0$) determine the kinetics of accumulation (segregation) of the interstitial element in the vicinity of the interface as a result of the chemical interaction (Fig. 1). The interstitial element is accumulated in the contact layer on defects modeled as "traps" for the diffusant. Then its concentration in the surface layer in the bound state (in nitride or oxide compounds) and its total concentration (in the solid solution and compounds) in the vicinity of the surface are computed from the relations

$$C^*(0,\tau)=\left(\frac{k}{\omega}\right)\int_0^\tau [C(0,t)-C_0]dt, \qquad C_\Sigma(0,\tau)=C(0,\tau)+C^*(0,\tau). \tag{6}$$

Thus, only a part of all adsorbed interstitial element atoms dissolves in the metal and diffuses in the volume. The remaining nitrogen atoms segregate in the form of compounds near the surface (Fig. 3).

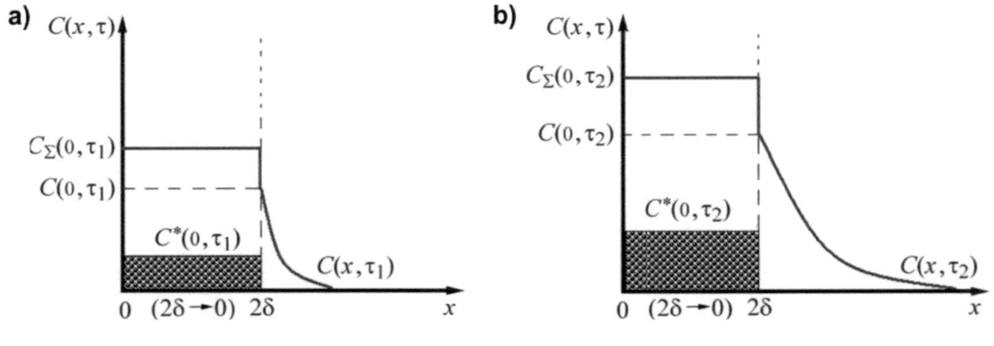

Figure 3. Evolution of the interstitial element distribution during diffusive saturation: (a: $\tau=\tau_1$, b: $\tau=\tau_2$; $\tau_1<\tau_2$

The evolution of spatial distribution of dissolved nitrogen in titanium (Fig. 3) during its thermodiffusive saturation gives a solution of the equations (1), (2), which in the analytic form is as follows (Matychak et al., 2007):

$$\bar{C}(x,\tau) = h(h+k)^{-1} erfc\left[x/(2\sqrt{D\tau})\right] - h\left[q_2^{-1}F_2(x,\tau) - q_1^{-1}F_1(x,\tau)\right]/(D\Delta), \tag{7}$$

where

$$F_1(x,\tau) = \exp(q_1 x + q_1^2 D\tau)\cdot erfc[q_1\sqrt{D\tau} + x/(2\sqrt{D\tau})],$$

$$F_2(x,\tau) = \exp(q_2 x + q_2^2 D\tau)\cdot erfc[q_2\sqrt{D\tau} + x/(2\sqrt{D\tau})],$$

$$q_1 = (1+\Delta)/2\omega, \quad q_2 = (1-\Delta)/2\omega, \quad \Delta = \sqrt{1-4\omega(h+k)/D}.$$

Specifically, the surface concentration of dissolved interstitial element is

$$\bar{C}(0,\tau) = h/(h+k) - [f_2(\tau)/q_2 - f_1(\tau)/q_1]\cdot h/(D\Delta), \tag{8}$$

where

$$f_1(\tau) = \exp(q_1^2 D\tau)\cdot erfc(q_1\sqrt{D\tau}), \quad f_2(\tau) = \exp(q_2^2 D\tau)\cdot erfc(q_2\sqrt{D\tau}). \tag{9}$$

Here $\bar{C}(x,\tau) = (C(x,\tau) - C_0)/(C_{eq} - C_0)$ is the relative change of the interstitial element concentration in the solid solution in α-titanium. Its surface concentration $C^*(0,\tau)$ in the bound state and the total concentration $C_\Sigma(0,\tau)$ are determined by formulas (6).

The obtained results for the diffusive saturation of titanium with nitrogen under low partial pressure (1 Pa) in the temperature range of 750-850 ^{0}C (below the temperature of allotropic transformation) were confirmed by the experimental results (Matychak et al., 2009).

2.2. Technique and results of experimental tests

2.2.1. Methods

The samples (10×15×1 mm) of VT1-0 commercially pure titanium were investigated after an isothermal exposure at temperatures of 750, 800, and 850 ^{0}C for 1, 5, and 10 h in a rarefied (to 1 Pa) dynamic nitrogen atmosphere (the specific inleakage rate was 7×10^{-3} Pa/sec). Before treatment, samples were ground to $R_a = 0{,}4$ μm, washed in acetone and alcohol, and dried.

Upon loading the samples in an ampoule, the system was pumped down to a pressure of 10^{-3} Pa, then the nitrogen was blown through, and required parameters of the gas medium were set. Heating was performed at a rate 0.04 ^{0}C/sec. After an isothermal exposure, the samples were furnace-cooled in nitrogen (the mean cooling rate was 100 ^{0}C/h).

Commercially-pure gaseous nitrogen was used, which, according to a technical specification, contained not more than 0.4 vol. % of oxygen and 0.07 g/m^3 of water vapor. Before feeding in the reaction space of a furnace, nitrogen was purified from oxygen and moisture by passing through a capsule with silica gel and titanium chips heated to a temperature higher by 50 ^{0}C than the saturation temperature. After every 3–4 tests, to restore the efficiency of the system for purification of nitrogen, silica gel was annealed at 180 ^{0}C for 3–4 h, and titanium chips were replaced by new ones. Due to this, the oxygen concentration in nitrogen ranged from 0.01 to 0.03 vol. %.

The microstructure of "oblique" microsections of samples was studied with a "Epiquant" microscope equipped with a camera and a computer with digital image analysis software.

The surface hardening was assessed based on the microhardness measured with a PMT-3M unit under a load of 0.49 N. As the depth of a nitrided layer, the depth of a zone was accepted in which the microhardness was higher than that of the core by δH=0.2 GPa (Fedirko & Pohrelyuk, 1995).

2.2.2. Results of experimental investigations

An analysis of experimental data of the influence of the partial nitrogen pressure on the saturation of titanium alloys during nitriding indicates that, in the range of rarefaction of the active gas 0.1–10 Pa (the specific inleakage rate ranged from 7×10^{-2} to 7×10^{-4} Pa/sec), the kinetics of nitriding is sensitive to processes related to the nitrogen feed to the gas–metal interaction zone (Fedirko & Pohrelyuk, 1995). Under such conditions, in a certain time range, which depends on the nitriding temperature, one can maintain the dynamic equilibrium between the adsorbed nitrogen and nitrogen transported by diffusion in the depth of the titanium matrix and shift significantly in time the beginning of the formation of a continuous nitride film. Metallographic analysis of the surface of VT1-0 titanium samples nitrided in this range of gas-dynamic (1 Pa; 7×10^{-3} Pa/sec) and temperature–time (750-850 ^{0}C; 5 h) parameters confirmed the absence of a continuous nitride film on their surfaces (Fig. 4). Instead of it, we observe the initiation and growth of nitride islands, predominantly between grains (Fig. 4 a-c).

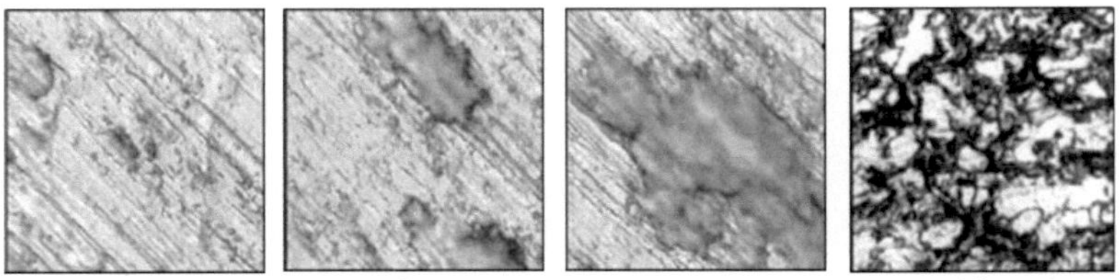

(a) (b) (c) (d)

Figure 4. Surface of VT1-0 titanium after nitriding in a rarefied dynamic nitrogen atmosphere (1 Pa): (a) 750 ^{0}C, 5 h; (b) 800 ^{0}C, 5 h;(c) 850 ^{0}C, 5 h; (d) 850 ^{0}C, 10 h

After an exposure for 10 h at 850 ^{0}C (Fig. 4 d), almost all grain boundaries contain nitrides, which favor the formation of a surface network from nitride inclusions and the formation of the corresponding surface topography. The dissolution of nitrogen in titanium stabilizes an

α-solid solution, the layer of which increases in thickness as the temperature–time parameters increase; its grains are etched less than the matrix (Fig. 5).

The surface microhardness of titanium changes as a result of nitride formation. After nitriding at 750 and 800 ^{0}C for 5 h and at 850 ^{0}C for 1 h when nitride formation is not very intensive, which is evidenced by reflexes of relative intensity of the nitride of lower valence (Ti_2N), the surface microhardness of titanium ranges from 4.4 to 7.3 GPa (Fig. 6). As the exposure time increases to 5 – 10 h at 850 ^{0}C when nitride islands cover a major part of the surface of the alloy, it rises to 10–13 GPa. The temperature–time nitriding parameters affect the surface microhardness of titanium and the depth of the nitrided layer, which increases monotonically in thickness with the temperature and time of exposure in a nitrogen-containing atmosphere (Fig. 6 c, d). Temperature influences analogously the surface microhardness for a given exposure time (Fig. 6 a). The effect of the time of saturation at 850 ^{0}C is somewhat different. As the exposure time increases from 1 to 5 h, the microhardness increases 2.5 times, and as the exposure time increases from 5 to 10 h, it rises only by 1.67 GPa (Fig. 6 b).

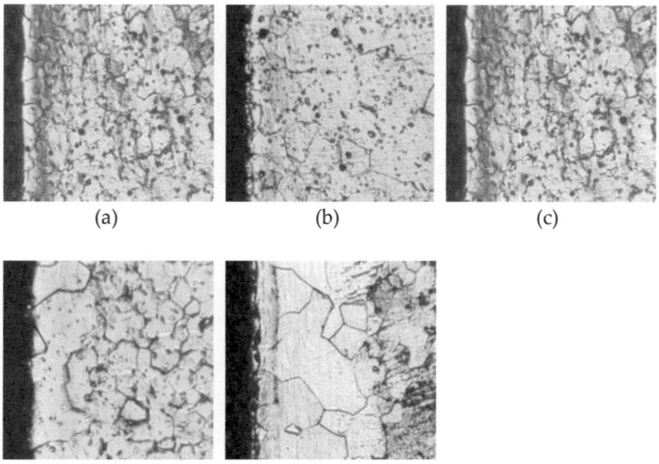

Figure 5. Microstructure of surface layers of VT1-0 titanium nitrided in a rarefied dynamic nitrogen atmosphere (1 Pa): (a) 750 ^{0}C, 5 h;(b) 800 ^{0}C, 5 h; (c) 850 ^{0}C, 1 h; (d) 850 ^{0}C, 5 h; (e) 850 ^{0}C, 10 h

Curves of the distribution of the microhardness over a cross-section of the hardened surface layers shift in the direction of higher values of the hardness with increases in the saturation temperature (Fig. 7 a) and saturation time (Fig. 7 b). During nitriding at a temperature of 850 ^{0}C for 5 h, the surface hardening of titanium is more significant than those at 750 and 800 ^{0}C and the same exposure time (Fig. 7 a).

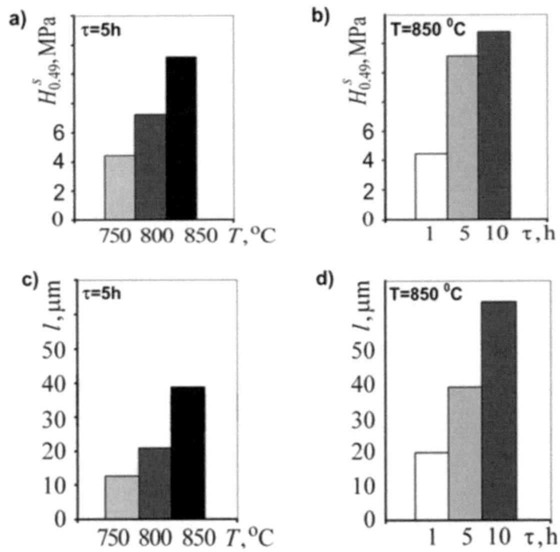

Figure 6. Dependences of the surface microhardness (a, b) and the depth of the hardened zone (c, d) on the temperature–time nitriding parameters of VT1-0 titanium in a rarefied dynamic atmosphere (1 Pa)

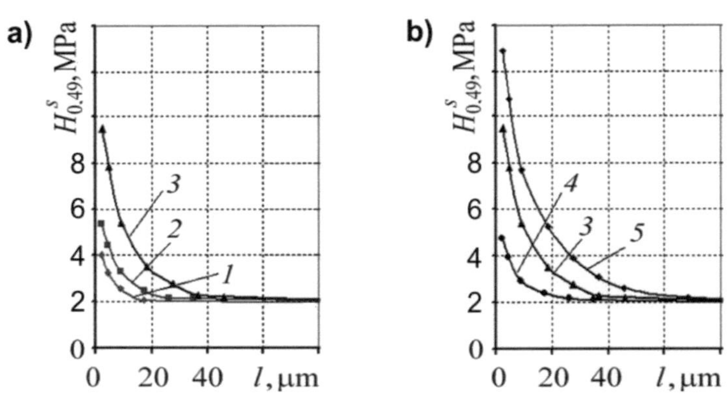

Figure 7. Distribution of the microhardness over the cross-section of surface layers of VT1-0 titanium nitrided in a rarefied dynamic atmosphere (1 Pa) depending on the temperature (a) and time of an isothermal exposure (b): (1) 750 °C, 5 h; (2) 800 °C, 5 h; (3) 850 °C, 5 h; (4) 850 °C, 1 h; (5) 850 °C, 10 h

2.3. Assessment of the temperature–time parameters of nitriding and analysis of results

It is known that the profile of nitrogen concentration in the surface layer of titanium substantially affects its physicomechanical properties. For experimental investigations of hardened nitrided layers, the method of layer-by layer testing of microhardness, which substantially depends on the content of dissolved nitrogen in titanium, is widely used. Let us use a known linear dependence of change of the microhardness on the concentration of an interstitial impurities in titanium (Korotaev et al., 1989):

$$H(x,\tau) = H_0 + a\cdot[C(x,\tau) - C_0] \tag{10}$$

Then the relative change in the microhardness in the diffusion zone due to dissolved nitrogen (neglecting the contribution of nitride inclusions) is as follows:

$$\bar{H}(x,\tau) = [H(x,\tau) - H_0]/[H_{max} - H_0] \equiv \bar{C}(x,\tau) \tag{11}$$

Here H_0 is the microhardness of the initial titanium sample, H_{max} is the microhardness of titanium at a maximum concentration of dissolved nitrogen $C_{max}=C_{eq}$, and a is the proportionality coefficient. Relation (11) indicates the possibility to plot the calculated relative concentrations $\bar{C}(x, \tau)$ of nitrogen and experimental data of the relative change in the microhardness $\bar{H}(x, \tau)$, on the same ordinate axis.

The roles of the time parameter and temperature are illustrated by analytic curves of the nitrogen content (Fig. 8, Fig. 9, Fig. 10), constructed from relations (6) – (8), and experimental data (see Fig. 4) using relation (11). For analytic calculations the following parameters were used: for T=750 ^{0}C – D =1•10^{-11} cm^2/sec, h=1•10^{-8} cm^2/ sec; for T=800 ^{0}C – D = 3.4•10^{-11} cm^2/sec, h = 3•10^{-8} cm/sec; for T = 850 ^{0}C – D=1•10^{-10} cm^2/sec, h=1•10^{-7} cm/sec; ω = 10^{-5} cm, k/h = 0.002. Diffusion coefficients of nitrogen were calculated according to the dependence $D = D_0 \exp(-E / RT)$, where D_0=0.96 cm^2/sec, E=214.7 kJ/mole (Metin & Inal, 1989). An analysis of these curves gives grounds for the following conclusions.

For an isothermal exposure T=850 ^{0}C in nitrogen, with increase in the saturation time, both the surface concentration of dissolved nitrogen C(0, τ) (curve 1) and its concentration in nitride inclusions $C^*(0, \tau)$ (curve 2), as well as its total content $C_\Sigma(0, \tau)$ (curve 3, Fig. 8 a), increase.

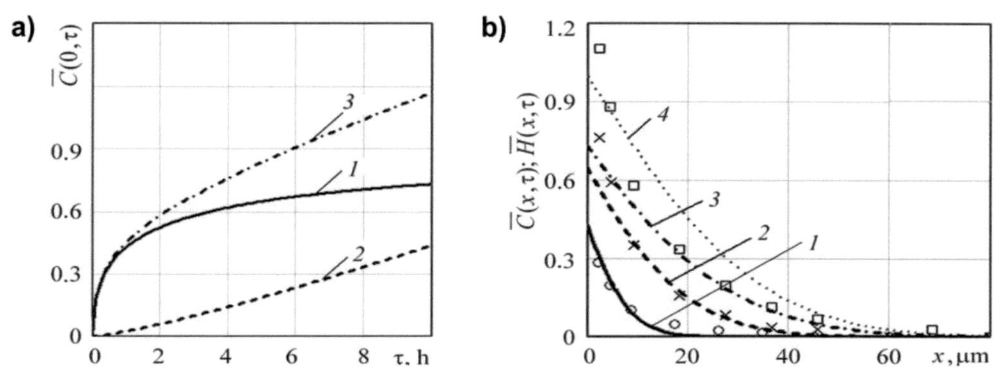

Figure 8. Time dependence of the surface nitrogen content ((a): curve (1) in the solid solution$C(0, \tau)$; (2) in nitride inclusions $C^*(0, \tau)$; (3) total content $C_\Sigma(0, \tau)$and its distribution in the surface layer of titanium (b) for different exposure times: (1), τ = 1 h; (2) τ = 5 h, (3) and (4) τ = 10 h (curve (4) was constructed for the condition $C(0, \tau)$=const,$h \to \infty$) at a nitriding temperature T = 850 ^{0}C. Marks correspond to experimental data (H(x, τ))

The nitrogen content in the surface layer and the depth of the diffusion zone change additively as the exposure time (curves 1 – 3, Fig. 8 b) and the temperature of the isothermal exposure (curves 1 – 3, Fig. 9) change.

On the whole, the analytic calculations of content profiles correlate well with the experimental results of relative changes in the microhardness of the surface layer (Fig. 8 b, Fig. 9). The corresponding curves have a monotonic character; the microhardness over the cross-section of the sample decreases gradually in the depth of the metal until it attains values characteristic for titanium. At the same time, there are insignificant disagreements between the theoretical and experimental results. In particular, for short exposures, the zone of change of the microhardness extends to a larger depth than the value which follows from the nitrogen distribution (Fig. 8 b). This can be explained by an insignificant content of oxygen, which is characterized by a larger diffusion mobility than nitrogen. For larger exposure times when the percentage of nitrogen is larger than that of oxygen, this effect is leveled.

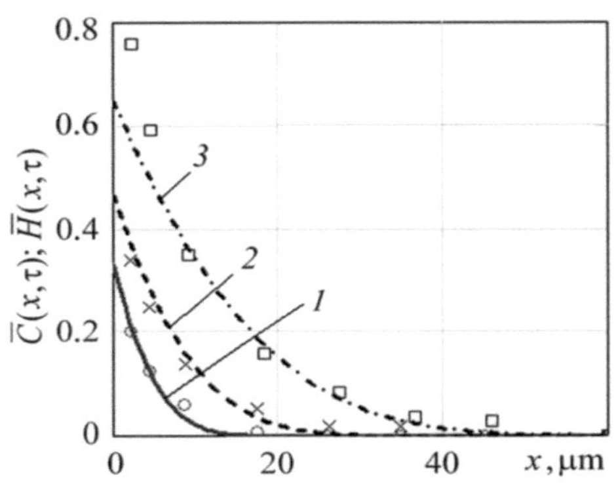

Figure 9. Distribution of nitrogen in the surface layer of titanium after an exposure τ= 5 h for different saturation temperature: (1), 750 ^{0}C; (2), 800 ^{0}C; (3), 850 ^{0}C. Marks correspond to experimental data $\bar{H}(x, \tau)$

Some disagreement between the calculated nitrogen distribution and experimental data of change in the microhardness is also observed near the surface, particularly as the time (Fig. 8 b) and temperature (Fig. 9) of the treatment increase. In our opinion, this is due to the influence of nitride inclusions, the content of which increases under such conditions, on the microhardness. That is why it is more expedient to use the modified dependence (11) with allowance for such an influence.

Not only data of the surface concentration of nitrogen (correspondingly, the hardness as well), but also data of its concentration at a certain distance from the surface and the depth of the nitrided layer depending on the temperature–time parameters are of practical interest. The corresponding curves (Fig. 10) were constructed for the same parameters as in the preceding figures. It should be noted that the depth of the diffusion zone was determined behind the front of propagation of the relative nitrogen concentration $\bar{C}$=0.02, which corresponds to a change in the microhardness by an amount H_δ = 0.2 GPa, equal to the error in its measurements.

An analysis of these curves confirms an adequate increase in the nitrogen content over the whole depth of the diffusion zone and the increase in the depth of this zone as the treatment time and temperature increase (curves 1–3, Fig. 10). It was found that, in the statement of the first boundary-value problem ($h \rightarrow 0$, $C(0, \tau) = Const$), overestimated values of the nitrogen concentration were obtained (curves 4 in Fig. 8 b Fig. 10). The calculated data show that the surface phenomena affect substantially not only the surface concentration of nitrogen (Fig. 10 a) (correspondingly, the surface hardness of titanium), but also the nitrogen content in layers more remote from the surface (Fig. 10 b), and the depth of the hardened diffusion zone (Fig. 10 c).

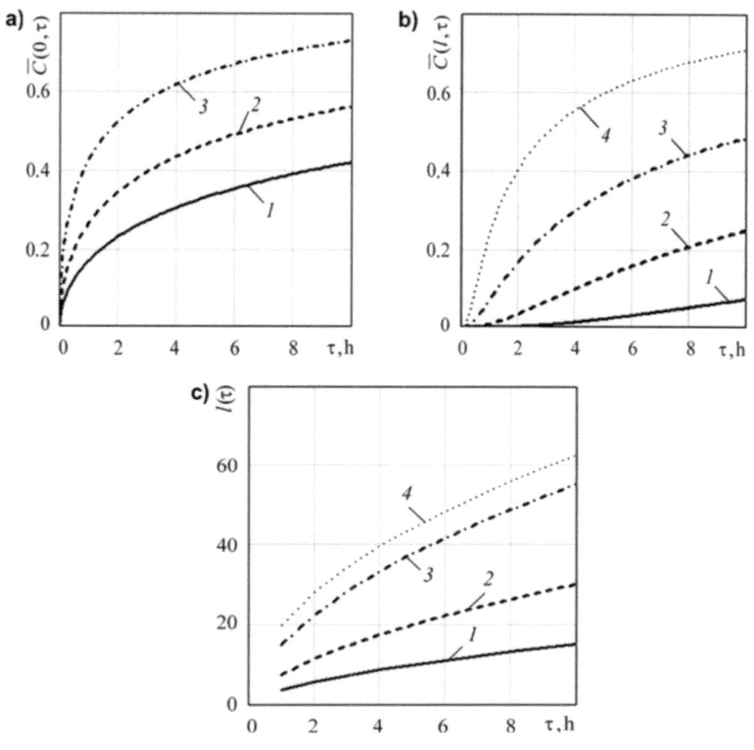

Figure 10. Time dependence of the content of dissolved nitrogen on the surface (a) and at a depth x = 10 μm (b), and the depth of the nitrided layer (c) for different saturation temperatures: (1) 750 °C; (2) 800 °C; (3, 4) 850 °C (curves 4 were constructed for the condition $C(0, \tau)$ = const, h → ∞)

Thus, the presented results indicate the critical role of the surface phenomena (adsorption and chemisorption) in the kinetic regularities of nitriding of titanium in a rarefied atmosphere. The calculated data obtained on the basis of the solution of the diffusion task using the nonstationary boundary condition (2) indicate that its model representation reflects rather satisfactorily the main tendencies of the high-temperature interaction of titanium with rarefied nitrogen. For the provision of a specified hardened layer, the proposed model gives scientifically justified recommendations on external parameters (exposure temperature and time) of nitriding of titanium.

3. Kinetic peculiarities of thermodiffusion saturation of titanium with interstitial elements at $T>T_{\alpha\leftrightarrow\beta}$

3.1. Thermodynamic analysis

According to the phase diagram (Fig. 11 a, b), titanium undergoes allotropic transformation (change of crystal lattice from hcp to bcc) at $T_{\alpha\leftrightarrow\beta}$ = 882 ⁰C (Fromm & Gebhardt, 1976).

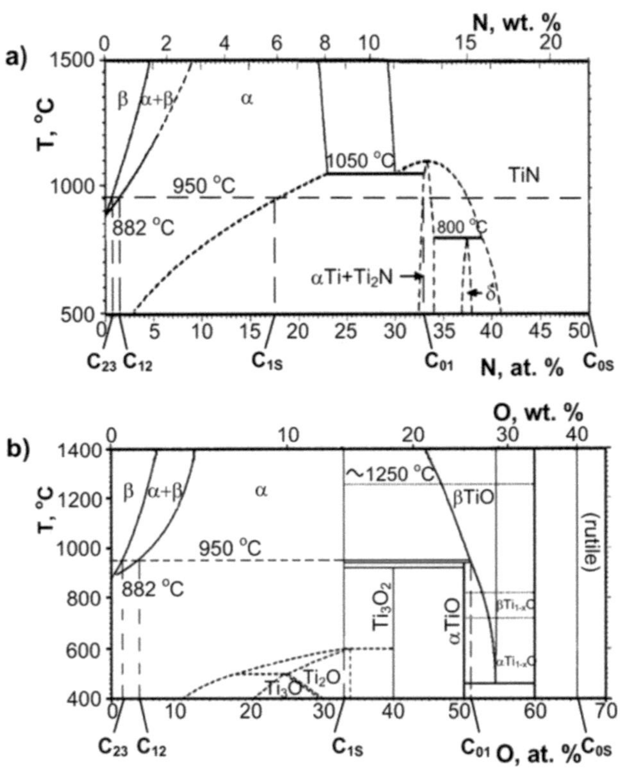

Figure 11. Ti-N (a) and Ti-O (b) phase diagrams (Fromm & Gebhardt, 1976)

We will be interested in high-temperature ($T>T_{\alpha\leftrightarrow\beta}$) interaction of titanium with the interstitial element A (A – N (nitrogen) or O (oxygen)). Under these conditions, according to the phase diagrams (Fig. 11 a, b), titanium nitrides or oxides (TiA_x) as products of chemical reactions and solid solutions of nitrogen or oxygen in α and β-phases of titanium are stable in the system.

In particular, in the concentration range $0<C_A<C_{23}$ solid solution of interstitial element in β-phase is stable, while in the concentration range $C_{12}<C_A<C_{1S}$ – solid solution of interstitial element in α-phase. In the concentration range $C_{23}<C_A<C_{12}$ solid solutions of interstitial element in α- and β-phases can coexist.

It should be noted that the solubility of nitrogen and oxygen in α-phase is high in comparison with β-phase. At the same time, their diffusion coefficients in α-phase are by two orders lesser than in β-phase (Fedirko & Pohrelyuk, 1995; Fromm & Gebhardt, 1976; Panasyuk, 2007). The solubility and diffusion coefficient of oxygen in α- and β-phases are much higher in comparison with nitrogen.

3.2. Physico-mathematical model

Let us consider the process of isothermal saturation of titanium by nitrogen or oxygen at temperature higher than temperature of allotropic transformation ($T>T_{\alpha\leftrightarrow\beta}$). In this case the initial microstructure of titanium consists of β-phase. According to the thermodynamic analysis, the following scheme of the gas-saturated layer of titanium is suggested (Fig. 12) (Tkachuk, 2012).

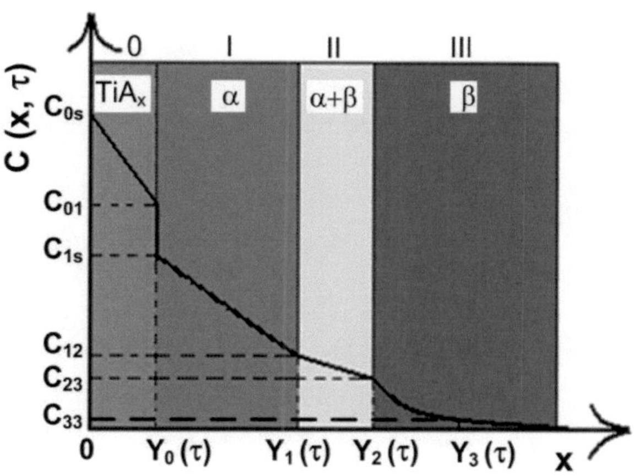

Figure 12. Scheme of the concentration distribution of interstitial element A (N or O) during saturation of titanium at $T>T_{\alpha\leftrightarrow\beta}$

During the interaction of titanium with nitrogen or oxygen nitride or oxide layer ($0 < x < Y_0(\tau)$) and diffusion zone are formed. The diffusion zone consists of three layers. The layer I ($Y_0(\tau) < x < Y_1(\tau)$), which borders on the nitride layer, is α-phase, significantly enriched in nitrogen or oxygen because of their high solubility in α-phase. This layer is formed and it grows during saturation because of diffusion dissolution of nitrogen or oxygen and structural transformations in titanium, because these interstitial elements are α-stabilizers. The layer III ($Y_2(\tau) < x < \infty$), which borders on the titanium matrix, at the temperature of saturation consists of β-phase enriched by nitrogen or oxygen. Between the first and third layers the layer II ($Y_1(\tau) < x < Y_2(\tau)$) is formed, which is the dispersed mixture of α- and β-phases, enriched in nitrogen or oxygen.

For analytical description of the process of saturation of titanium by nitrogen or oxygen some model assumptions should be done. The aim of thermochemical treatment of titanium samples is strengthening of their surface layer and as the object of analytical investigation of the kinetics of diffusion saturation of titanium the half-space ($0\leq x<\infty$) has been chosen. Nitride or oxide film is formed immediately. Surface concentration of nitrogen or oxygen does not change with time and corresponds to stoichiometric titanium nitride (TiN) or oxide (TiO_2). On the interfaces the nitrogen or oxygen concentration, corresponding to equilibrium concentration, according to the phase diagram is constant (Fig. 11 a, b).

The diffusion process in such heterogeneous system will be described by Fick's system of equations:

$$D_i \partial^2 C_i(x,\tau) / \partial x^2 = \partial C_i(x,\tau) / \partial \tau \ , \quad i = 0,1,2,3. \qquad (12)$$

Here $C_i(x,\tau)$ and D_i are concentration and diffusion coefficients of the interstitial element; index i=0 corresponds to nitride TiN_x or oxide TiO_{2-x} layer ($0 < x < Y_0(\tau)$); i=1 – α-Ti layer ($Y_0(\tau) < x < Y_1(\tau)$); i=2 – (α+β)-Ti layer ($Y_1(\tau) < x < Y_2(\tau)$); i=3 – β-Ti layer ($Y_2(\tau) < x < \infty$)).

Initial conditions (τ=0):

$$C_i(x,0) = 0, \quad Y_i(0) = 0 \quad for\ x > 0. \qquad (13)$$

Boundary conditions (τ>0):

$$\begin{gathered} C_0(0,\tau) = C_{0S}, \quad C_3(\infty,\tau) = 0, \qquad C_0[Y_0(\tau),\tau] = C_{01}, \quad C_1[Y_0(\tau),\tau] = C_{1S}, \\ C_1[Y_1(\tau),\tau] = C_2[Y_1(\tau),\tau] = C_{12}, \quad C_2[Y_2(\tau),\tau] = C_3[Y_2(\tau),\tau] = C_{23}. \end{gathered} \qquad (14)$$

The motion of interfaces will be set by the parabolic dependencies (Lyubov, 1981):

$$Y_0(\tau) = 2 \cdot \beta_0 \cdot \sqrt{D_0 \cdot \tau}, \quad Y_1(\tau) = 2 \cdot \beta_1 \cdot \sqrt{D_1 \cdot \tau}, \quad Y_2(\tau) = 2 \cdot \beta_2 \cdot \sqrt{D_2 \cdot \tau}. \qquad (15)$$

Here β_j (j=0,1,2) are dimensionless constants (for the specific temperature), which will be determined from the law of conservation of mass on the interfaces. Thus for diffusion fluxes on the interfaces $Y_j(\tau)$are set:

$$\begin{gathered} -D_0 \left.\frac{\partial C_0}{\partial x}\right|_{x=Y_0(\tau)-0} + D_1 \left.\frac{\partial C_1}{\partial x}\right|_{x=Y_0(\tau)+0} = (C_{01} - C_{1S}) \frac{dY_0(\tau)}{d\tau}, \\ D_1 \left.\frac{\partial C_1}{\partial x}\right|_{x=Y_1(\tau)-0} = D_2 \left.\frac{\partial C_2}{\partial x}\right|_{x=Y_1(\tau)+0}, \quad D_2 \left.\frac{\partial C_2}{\partial x}\right|_{x=Y_2(\tau)-0} = D_3 \left.\frac{\partial C_3}{\partial x}\right|_{x=Y_2(\tau)+0} \end{gathered} \qquad (16)$$

It is difficult to solve the equations system (12) – (16) in analytical form. The method of approximate solution of above mentioned task should be used (Lykov, 1966). It is accepted the linear distribution law of the concentration of the interstitial element in TiA_x layer (Fig. 12) corresponded to the quasi-stationary state. It is accepted the same distribution law in the first two layers of diffusion zone. It was considered that in the third layer of diffusion zone the distribution of the interstitial element is realized by Gauss's law:

$$C_0(x,\tau)=C_{0S}-(C_{0S}-C_{01})\frac{x}{Y_0(\tau)},\quad C_1(x,\tau)=C_{1S}-(C_{1S}-C_{12})\frac{x-Y_0(\tau)}{Y_1(\tau)-Y_0(\tau)},$$
$$C_2(x,\tau)=C_{12}-(C_{12}-C_{23})\frac{x-Y_1(\tau)}{Y_2(\tau)-Y_1(\tau)},\quad C_3(x,\tau)=C_{23}erfc\frac{x-Y_2(\tau)}{2\sqrt{D_3\tau}}. \tag{17}$$

The chosen functions $C_i(x,\tau)$ satisfy the initial (13) and boundary (14) conditions as well as the differential equations (12). The following system of equations for calculating the parameters β_j (j=0,1,2) was obtained by the conditions of mass balance on interfaces (16) and relation (15) (Tkachuk et al., 2012):

$$\frac{A_0}{2\beta_0}\left[\frac{1}{\beta_0}-\frac{1}{A_1\lambda_0(\beta_1-\beta_0\lambda_0)}\right]=1,\ \frac{(\beta_1-\beta_0\lambda_0)}{A_2\lambda_1(\beta_2-\beta_1\lambda_1)}=1,\ \frac{2(\beta_2-\beta_1\lambda_1)}{A_3\lambda_2\sqrt{\pi}}=1. \tag{18}$$

Having solved the system of equations (18), following equations were:

$$\beta_0=A_0(\sqrt{B^2+2/A_0}-B)/2,\qquad \beta_1=\beta_0\lambda_0+A_2A_3\lambda_1\lambda_2\sqrt{\pi}/2,$$
$$\beta_2=\beta_1\lambda_1+A_3\lambda_2\sqrt{\pi}/2, \tag{19}$$

where

$$B=1/[A_1A_2A_3\lambda_0\lambda_1\lambda_2\sqrt{\pi}],\quad \lambda_0=\sqrt{D_0/D_1},\quad \lambda_1=\sqrt{D_1/D_2},\quad \lambda_2=\sqrt{D_2/D_3},$$
$$A_0=\frac{C_{0S}-C_{01}}{C_{01}-C_{1S}},\quad A_1=\frac{C_{0S}-C_{01}}{C_{1S}-C_{12}},\quad A_2=\frac{C_{1S}-C_{12}}{C_{12}-C_{23}},\quad A_3=\frac{C_{12}-C_{23}}{C_{23}}.$$

The parameters β_j depend on concentration of nitrogen or oxygen on interfaces and their diffusion coefficients in α- and β-phases, which in turn depend on the temperature. In particular, for saturation temperature of T=950 ^{0}C the diffusion coefficients of nitrogen and oxygen in surface layers of titanium, and equilibrium concentrations of nitrogen and oxygen on interfaces, according to the corresponding phase diagrams (Fig. 11 a, b), are presented in Table 1.

Taking the values of these parameters, according to relations (19), the constants β_j for nitrogen and oxygen were calculated (Table 2).

A	D_0, cm^2/sec	D_1, cm^2/sec	D_2, cm^2/sec	D_3, cm^2/sec	C_{0S}, at. %	C_{01}, at. %	C_{1S}, at. %	C_{12}, at. %	C_{23}, at. %	C_{33}, at. %
N	3×10^{-12}	2.5×10^{-10}	2.5×10^{-9}	3.2×10^{-8}	50	33	17.5	1.5	0.75	0.25
O	2.5×10^{-11}	2.1×10^{-9}	2.1×10^{-8}	1.6×10^{-7}	66	51	33	4	2	0.25

Table 1. Diffusion coefficients of nitrogen and oxygen in the surface layer of titanium (Fedirko & Pohrelyuk, 1995; Fromm & Gebhardt, 1976; Panasyuk, 2007) and equilibrium concentration of nitrogen and oxygen on the interfaces at saturation temperature of T=950 ^{0}C

A	β_0	β_1	β_2	K_0, $cm/sec^{1/2}$	K_1, $cm/sec^{1/2}$	K_2, $cm/sec^{1/2}$
N	0.183	1.691	0.782	6.328×10^{-7}	5.348×10^{-5}	7.825×10^{-5}
O	0.082	1.481	0.789	8.175×10^{-7}	1.357×10^{-4}	2.288×10^{-4}

Table 2. Calculated constants β_j and K_j (j=0,1,2) at saturation temperature of T=950 ^{0}C

Taking into consideration the correlation (15), the motion of interfaces will be presented as:

$$Y_0(\tau) = K_0\sqrt{\tau}\ ,\quad Y_1(\tau) = K_1\sqrt{\tau}\ ,\quad Y_2(\tau) = K_2\sqrt{\tau}\ , \tag{20}$$

where $K_0=2\beta_0\sqrt{D_0}$, $K_1=2\beta_1\sqrt{D_1}$, $K_2=2\beta_2\sqrt{D_2}$ – constants of the parabolic growth of nitride or oxide layer and α, (α+β) layers of diffusion zone stabilized by nitrogen or oxygen. In particular, for saturation temperature of T=950 ^{0}C these calculated constants are presented in Table 2.

Having found the constants of parabolic growth of the layers, and having used the relations (20), it is easy to foresee the kinetics of motion of interfaces: $Y_0(\tau)$ – interface of nitride or oxide layer (Fig. 13 a), $Y_1(\tau)$ – interface of solid solution of interstitial element in α-phase (Fig. 13 b), $Y_2(\tau)$ – interface of mixture of solid solutions of interstitial element in α- and β-phases (Fig. 13 c), $Y_3(\tau)$ – interface of solid solution of interstitial element in β-phase (Fig. 13 d) at nitriding and oxidation of titanium at T=950 ^{0}C. One could notice that the last interface is identified by the motion of conventional boundary with the specific nitrogen or oxygen concentration, for example C_{33} = 0.25 at. %, that is from transcendental equation $C_3(Y_3(\tau), \tau)=C_{33}$.

Calculated constants $Y_i(\tau)$ (i=0,1,2,3) after isothermal exposures of 1 and 5 h during nitriding and oxidation of titanium at T=950 ^{0}C are presented in Table 3. It is clear that according to the assumptions (15) with the increase of processing time the motion of interfaces (Fig. 13 a, b, c, d) occur according to the parabolic dependences proportionally to the corresponding constants of parabolic growth K_j (j=0,1,2).

On the basis of relations (17) the concentration profiles of nitrogen (curves 1) and oxygen (curves 2) in the diffusion zone of titanium after nitriding and oxidation during 1 h (Fig. 14 a) and 5 h (Fig. 14 b) are calculated.

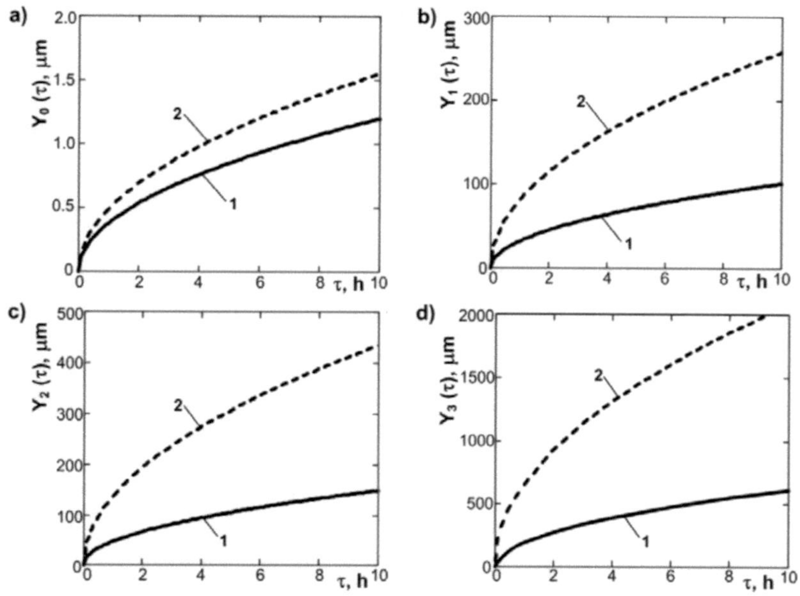

Figure 13. Kinetics of motion of interfaces $Y_0(\tau)$ (a), $Y_1(\tau)$ (b), $Y_2(\tau)$ (c) and $Y_3(\tau)$ (d) at nitriding (curves 1) and oxidation (curves 2) of titanium at saturation temperature of T=950 °C

The diffusion coefficient of nitrogen or oxygen in β-phase is by two-four orders higher than in α-phase and in nitride or oxide layers, that's why the thickness of β layer is much larger than the thickness of the other layers of diffusion zone (Fig. 13). If the thickness of nitride layer is less than 0.2% and oxide layer is less than 0.1 % of the total thickness of diffusion zone ($Y_3(\tau)$), the thickness of α, α + β and β layers will be 16, 8 and 76 % for nitriding and 12, 8 and 80 % for oxidation respectively.

	τ = 1 h				τ = 5 h			
A	Y_0, μm	Y_1, μm	Y_2, μm	Y_3, μm	Y_0, μm	Y_1, μm	Y_2, μm	Y_3, μm
N	0.4	32	47	194	0.85	72	105	433
O	0.5	81	137	658	1.1	182	307	1470

Table 3. Calculated $Y_i(\tau)$ (i=0,1,2,3) after isothermal exposures of 1 and 5 h at saturation temperature of T=950 °C

At the same time, the different solubility of nitrogen or oxygen in α and β-phases influences on the distribution of nitrogen or oxygen in the diffusion zone. When the structural phase transformations did not occur in the diffusion zone, the profiles of nitrogen and oxygen in this zone would be with a small gradients because of the low solubility of nitrogen or oxygen in β-phase. In fact, nitrogen and oxygen, being α-stabilizers, stimulate the β→ α phase transformation in the layers of the diffusion zone adjacent to nitride or oxide layer. And as the solubility of nitrogen and oxygen in α-phase is much higher than in β-phase, it can be foreseen that in zone I the profiles of nitrogen and oxygen will have a large gradients (Fig. 14), and respectively the distributions of microhardness in this zone will have a large gradients. It has been confirmed by the literature data (Lazarev et al., 1985) and the experimental investigations' data on nitriding.

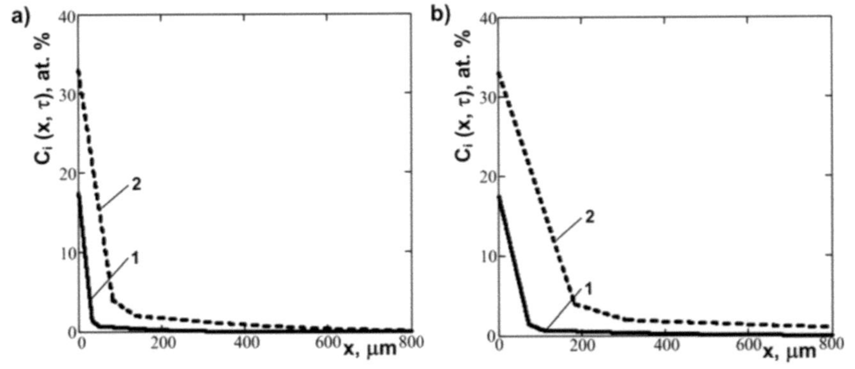

Figure 14. Concentration profiles of nitrogen and oxygen in diffusion zone of titanium after its saturation at T=950 ^{0}C for two isothermal exposures: a - τ = 1 h; b - τ = 5 h; curves 1 – for nitrogen, curves 2 – for oxygen

It was observed 2.5-3.0 times larger thickness of all layers of diffusion zone after oxidation comparing to nitriding (Fig. 13 b, c, d) as a result of the higher on order diffusion coefficients of oxygen in α- and β-phases compared to the diffusion coefficients of nitrogen (Table 1). Also the larger concentration gradient of oxygen in the layer I adjacent to oxide layer than concentration gradient of nitrogen in the layer I adjacent to nitride layer was received (Fig. 14). It is caused by higher solubility of oxygen in comparison with nitrogen in α-phase (Table 1).

3.3. Experimental procedure

Experimental investigation on the example of nitriding of titanium at the temperature of T=950 ^{0}C was conducted to check the validity of the above elaborated model representations.

Commercially pure (c.p.) titanium samples with dimensions of 10×15×4 mm were investigated. The samples were polished (R_a=0,4 μm), washed with deionized water prior to the treatment. The samples were heated to nitriding temperature in a vacuum of 10^{-3} Pa. Then they were saturated with molecular nitrogen of the atmospheric pressure at temperature of 950 ^{0}C. The isothermal exposure time in nitrogen was 1 and 5 h. After isothermal exposure the samples were cooled in nitrogen to room temperature.

The microstructure of nitride layers was studied by the use of metallographic microscope "EPIQUANT". Distribution of microhardness on cross section of surface layers of c.p. titanium after nitriding was estimated measuring microhardness at loading of 0.49 N.

3.4. Results and discussion

The nitride layer of goldish colour is formed on the surface of c.p. titanium after nitriding. Its colour is darkening with the increase of isothermal exposure time in nitrogen atmosphere. It indicates the increase of its thickness.

The diffusion zone is formed under titanium nitride layer (Fig. 15 a, b).

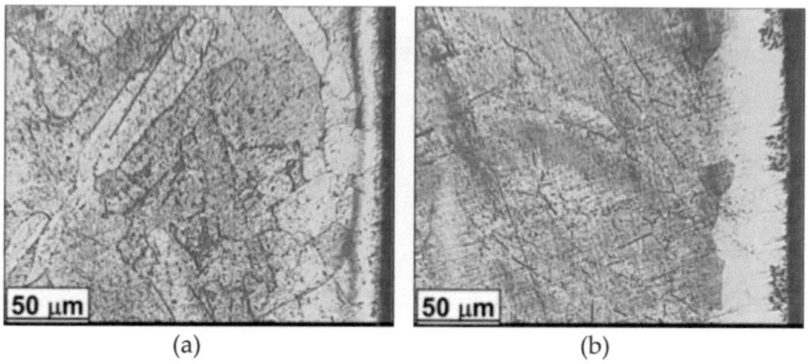

Figure 15. Microstructure of surface layer of c.p. titanium for nitriding at τ=1 (a) and 5 h (b) (T=950 °C, p=10^5 Pa)

It is difficult to find the layer II (Fig. 12) in this zone which, according to the phase diagram (Fig. 11 a) has to form. However, two parts of diffusion zone (zone A and zone B) of different structure are clearly identified. Zone A is α-phase formed during nitriding by nitrogen as α-stabilizer. Its thickness, according to data of metallographic analysis, increases from 20 to 45 μm with the increase of duration of nitriding from 1 to 5 h. Zone B is α-phase on the basis of solid solution of nitrogen, however formed as a result of β→ α transformation at cooling.

The results of investigation of character of distribution of microhardness on cross section of surface layers of c.p. titanium after nitriding are presented in Fig. 16 a. It is distinguished zone A (layer I, Fig. 12) and zone B (probably, layer II + layer III, Fig. 12) on curves of distribution of microhardness.

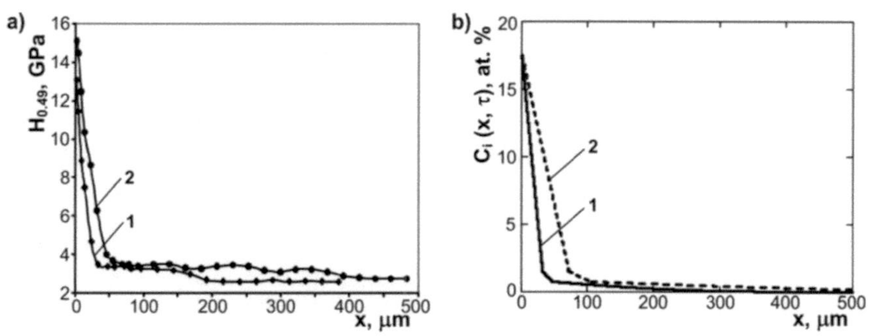

Figure 16. Distribution of nitrogen in diffusion zone of titanium (theory) (a) and distribution of microhardness on cross section of surface layer of c.p. titanium (experiment) (b) after its nitriding at T=950 °C and p=10^5 Pa for two isothermal exposure times: curve 1 – τ = 1 h; curve 2 – τ = 5 h

The large gradient of microhardness is characteristic for zone A. It is caused by $\beta \rightarrow \alpha$ transformation as a result of saturation by nitrogen as α-stabilizer and comparatively its high solubility in α-phase. With increasing distance from surface the microhardness is decreased sharply (Fig. 16 a) that is explained by decrease of nitrogen concentration (Fig. 16 b). The hardness of zone B is considerably less than zone A because of large difference of nitrogen solubility in α- and β-phases. The thickness of these zones is increased with the increase of duration of nitriding (Fig. 16 a). In particular, the thickness of zone A is 34 μm for τ= 1 h and 69 μm for τ=5 h. It can be noticed that this thickness is larger than corresponding thickness, determined by the data of metallographic analysis. The total depth of diffusion zone (zone A + zone B) is 185 μm for τ= 1 h and 425 μm for τ=5 h (Fig. 16 a).

The received analytical distribution of nitrogen (Fig. 16 b) and results of microhardness measurements (Fig. 16 a) confirm the correlation between model calculations and experimental data.

4. Conclusions

The process of high-temperature interaction of titanium with gaseous medium (nitrogen or oxygen) was modelled at temperatures $T<T_{\alpha\leftrightarrow\beta}$ and $T>T_{\alpha\leftrightarrow\beta}$ considering the surface processes and structural phase transformations.

The kinetics of surface processes is reflected by the mass balance equation, which takes into account the interaction of an external flux of impurities to the surface and its chemisorption with diffusion dissolution and segregation on defects as a result of a chemical interaction with titanium atoms.

The kinetics of diffusion saturation of α-titanium by nitrogen under rarefied atmosphere (1 Pa) in the temperature range of 750-850 ^{0}C was investigated experimentally and analytically. The influence of time and temperature parameters on the depth of the nitrided layer and a change of its microhardness was estimated.

It was shown the role of these interstitial elements as α-stabilizers in forming the diffusion zone which contains three layers based on α-phase, α+β-phases and β-phase.

It was received the solution of the formulated task as for diffusion of nitrogen or oxygen in such a heterogeneous medium taking into account the motion of interfaces.

The constants of parabolic growth of layers were calculated. It allowed to foresee the kinetics of their growth and distribution of interstitial elements (nitrogen or oxygen) in diffusion zone.

The adequacy of the proposed model representations was confirmed by the results of experimental investigations on nitriding of titanium at T=950 ^{0}C. The microstructural evolution (after processing times of 1 and 5 h) of the diffusion zone which is caused by the structural phase transformations during diffusion of nitrogen was examined experimentally.

Author details

Yaroslav Matychak*, Iryna Pohrelyuk, Viktor Fedirko and Oleh Tkachuk

Karpenko Physico-Mechanical Institute of National Academy of Sciences of Ukraine, Lviv, Ukraine

References

[1] Fedirko, V. M. & Pohrelyuk, I. M. (1995). *Nitriding of Titanium and its Alloys*, Naukova Dumka, Kiev

[2] Fedirko, V. M., Matychak, Ya. S., Pohrelyuk, I. M. & Prytula, A. O. (2005). Description of the diffusion saturation of titanium with nonmetallic admixtures with regard for their segregation on the surface. *Materials Science*, Vol. 41, No. 1, pp. 39–46

[3] Fromm, E. & Gebhardt, E. (1976). *Gase und Kohlenstoff in Metallen*, Springer-Verlag, Berlin-Heidelberg-New York

[4] Kofstad, P. (1966). *High-Temperature Oxidation of Metals*, John Wiley & Sons, New York

[5] Korotaev, A. D., Tyumentsev, A. N. & Sukhovarov, V. F. (1989). *Disperse Hardening of Refractory Metals*, Nauka, Novosibirsk

[6] Lazarev, E.M., Kornilova, Z.I. & Fedorchuk, N.M. (1985). *Oxidation of Titanium Alloys*, Nauka, Moskow

[7] Lykov, A.V. (1966). *Theory of Heat Conduction*, Vysshaya shkola, Moskow

[8] Lyubov, B.Ya. (1981). *Diffusive processes in heterogeneous solids*, Nauka, Moskow

[9] Malinov, S., Zhecheva, A. & Sha, W. (2003). Modelling the nitriding in titanium alloys. *ASM International*, pp. 344–352

[10] Matychak, Ya. S., Fedirko, V. M., Pavlyna, V.S. & Yeliseyeva, O.I. (1999). The research of initial stage of oxide diffusion growth in Fe-Pb-O system. *Metallofizika i Noveishie Tekhnologii*, Vol. 21 № 2, pp. 78-83

[11] Matychak, Ya., Fedirko, V., Prytula, A. & Pohrelyuk, I. (2007). Modeling of diffusion saturation of titanium by interstitial elements under rarefied atmospheres. *Defect Diffusion Forum*, Vol. 261–262, pp. 47–54

[12] Matychak, Ya. S., Pohrelyuk, I. M. & Fedirko, V. M. (2009). Thermodiffusion saturation of α-titanium with nitrogen from a rarefied atmosphere. *Materials Science*, Vol. 45, No. 1, pp. 72-83

[13] Metin, E. & Inal, O. T. (1989). Kinetics of layer growth and multiphase diffusion in ion-nitrided titanium. *Metallurgical and Materials Transactions A*, Vol. 20A, pp. 1819–1832

[14] Panasyuk, V.V. (2007). *Strength and Durability of Airplane Materials and Structural Elements*, Spolom, Lviv

[15] Prytula, A. O., Fedirko, V. M., Pohrelyuk, I. M. & Matychak, Ya. S. (2005). Surface chemical reactions in processes of diffusion mass transfer. *Defect Diffusion Forum*, Vol. 237–240, pp. 1312–1318

[16] Raichenko, A. I. (1981). *Mathematical Theory of Diffusion in Supplements*, Naukova Dumka, Kiev

[17] Tkachuk, O., Matychak, Ya., Pohrelyuk, I. & Fedirko, V. (2012). Diffusion of nitrogen and phase-structural transformations in titanium, *Proceedings of International Workshop "Diffusion, Stress, Segregation and Reactions (DSSR-2012)"*, Svitanok, Cherkasy region, June, 2012

Chapter 4

Microstructure and Mechanical Properties of High Strength Two-Phase Titanium Alloys

J. Sieniawski, W. Ziaja, K. Kubiak and M. Motyka

Additional information is available at the end of the chapter

http://dx.doi.org/10.5772/56197

1. Introduction

Two-phase titanium alloys constitute very important group of structural materials used in aerospace applications [1-3]. Microstructure of these alloys can be varied significantly in the processes of plastic working and heat treatment allowing for fitting their mechanical properties including fatigue behaviour to the specific requirements [4-6].

The main types of microstructure are (1) lamellar – formed after slow cooling when deformation or heat treatment takes place at a temperature in the single-phase β-field above the so-called beta-transus temperature T_β (at which the $\alpha+\beta \rightarrow \beta$ transformation takes place), consisting of colonies of hexagonal close packed (hcp) α-phase lamellae within large body centered cubic (bcc) β-phase grains of several hundred microns in diameter, and (2) equiaxed – formed after deformation in the two-phase $\alpha+\beta$ field (i.e., below T_β), consisting of globular α-phase dispersed in β-phase matrix [7-8].

The first type of microstructure is characterized by relatively low tensile ductility, moderate fatigue properties, and good creep and crack growth resistance.

The second microstructure has a better balance of strength and ductility at room temperature and fatigue properties which depend noticeably on the crystallographic texture of the hcp α-phase.

An advantageous balance of properties can be obtained by development of bimodal microstructure consisting of primary α-grains and fine lamellar α colonies within relatively small β-grains (10-20 μm in diameter) [9-10].

In the following sections the relations between microstructure morphology and mechanical properties of selected high strength two-phase titanium alloys were analysed.

Dilatometric tests, microstructure observation and X-ray structural analysis were carried out for cooling rates in the range of 48-0.004°C s^{-1} and time-temperature-transformation diagrams were developed for continuous cooling conditions (CCT).

The influence of the quantitative parameters of lamellar microstructure on the tensile properties and fatigue behaviour of selected two-phase titanium alloys was analysed. Rotational bending tests were carried out to determine high cycle fatigue (HCF) strength at 10^7 cycles.

2. High strength two-phase titanium alloys

The materials tested were high strength, two-phase $\alpha+\beta$ titanium alloys: Ti-6Al-4V, Ti-6Al-2Mo-2Cr and Ti-6Al-5Mo-5V-1Cr-1Fe (Table 1).

Alloy	Stability factor of β-phase	Alloying elements content, wt.%							
	K_β	Al	Mo	V	Cr	Fe	C	Si	Ti
Ti-6Al-4V	0.3	6.1	–	4.3	–	0.16	0.01	–	bal.
Ti-6Al-2Mo-2Cr	0.6	6.3	2.6	–	2.1	0.40	0.05	0.2	bal.
Ti-6Al-5Mo-5V-1Cr-1Fe	1.2	5.8	5.3	5.1	0.9	0.8	0.05	0.15	bal.

Table 1. Chemical composition of the investigated titanium alloys.

Ti-6Al-4V – martensitic $\alpha+\beta$ alloy ($K_\beta = 0.3$) – is the most widespread titanium alloy (>60% of all titanium alloys produced in USA and EU). Its high applicability results from good balance of mechanical properties and good castability, plastic workability, heat treatability and weldability. Aluminium addition stabilizes and strengthen α phase, increases $\alpha+\beta \leftrightarrow \beta$ transformation temperature and reduces alloy density. Vanadium – β-stabilizer – reduces $\alpha+\beta \leftrightarrow \beta$ transformation temperature and facilitates hot working (higher volume fraction of β-phase). Depending on required mechanical properties following heat treatment can be applied to Ti-6Al-4V alloy: partial annealing (600÷650ºC / 1h), full annealing (700÷850ºC / furnace cooling to 600ºC / air cooling) or solutioning (880÷950ºC / water quenching) and ageing (400÷600ºC) [1,3].

Ti-6Al-2Mo-2Cr – martensitic $\alpha+\beta$ alloy – known as VT3-1, is one of the first widespread high-temperature titanium alloys used in Russia for aircraft engine elements. Amount of β-stabilizers is similar to Ti-6Al-4V alloy but β-stabilizing factor is higher (K_β = 0.6). Mo – stabilises and strengthens β-phase, in the presence of Si increases creep resistance, facilitates plastic working, Cr, Fe – eutectoid elements, stabilise β-phase and strengthen α and β phases in the low and medium temperature range [1].

The alloy is processed by forging, stamping, rolling and pressing. Depending on the application and required properties following heat treatment can be applied to the semiproducts: isothermal annealing (870ºC / 1h / furnace cooling to 650ºC / holding for 2 h / air cooling),

http://dx.doi.org/10.5772/56197

duplex annealing (880ºC / 1h / air cooling and following heating 550ºC / 2÷5 h / air cooling) or hardening heat treatment (water quenching and ageing) [1,7].

Ti-6Al-2Mo-2Cr alloy retains its mechanical properties up to 300°C. At the temperature higher than 400°C mechanical properties are reduced due to partitioning of alloying elements proceeding by diffusion.

Ti-6Al-5Mo-5V-1Cr-1Fe transition α+β titanium alloy (K_β = 1.2) is produced in Russia and Ukraine, where is known as VT22. It is characterized by very good mechanical properties thus is mainly used for large, heavy loaded, forged parts for long-term operation at elevated temperature up to 350÷400ºC and short-term up to 750÷800ºC. Typical applications include disks and blades of low pressure compressors, landing gear elements, engine mount struts and others [1,11].

3. Development of microstructure during continuous cooling

Phase composition of titanium alloys after cooling from β phase range is controlled by cooling rate. Kinetics of phase transformations is related to the value of β-phase stability coefficient K_β resulting from the chemical composition of the alloy [7].

One important characteristic of the alloy is a range of α+β→β phase transformation temperature that determines conditions of thermomechanical processing intended for development of suitable microstructure. Start and finish temperatures of α+β→β phase transformation, vary depending on the contents of β stabilizing elements (Table 2).

Phase transformation		Alloy	
temperature, °C	Ti-6Al-4V	Ti-6Al-2Mo-2Cr	Ti-6Al-5Mo-5V-1Cr-1Fe
$T^{ns}_{\alpha+\beta\to\beta}$	890	840	790
$T^{ps}_{\alpha+\beta\to\beta}$	930	920	830
$T^{f}_{\alpha+\beta\to\beta}$	985	980	880
$T^{s}_{\beta\to\alpha+\beta}$	950	940	850
$T^{f}_{\beta\to\alpha+\beta}$	870	850	810

ns – nucleation start

ps – precipitation start

s – start

f – finish

Table 2. Start and finish temperature of the α+β→β phase transformation for selected titanium alloys ($v_h = v_c = 0.08°C\,s^{-1}$)

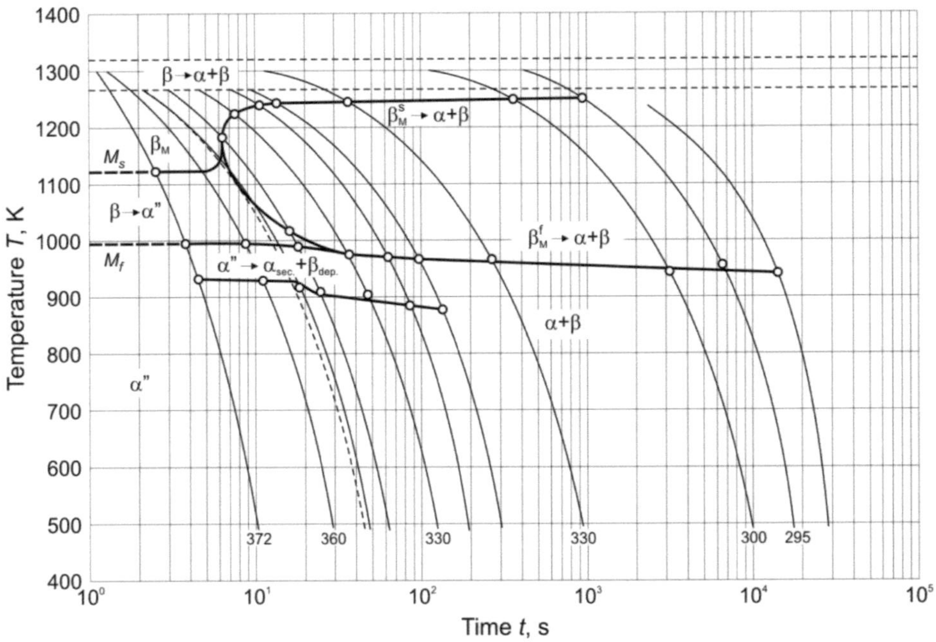

Figure 1. CCT diagram for Ti-6Al-4V alloy.

Cooling of Ti-6Al-2Mo-2Cr and Ti-6Al-4V alloys from above β transus temperature at the rate higher than 18°C s^{-1} leads to development of martensitic microstructure consisting of α'(α") phases (Fig. 4). Start and finish temperatures of the martensitic transformation β→α'(α") or β→α" do not depend on cooling rate but on β-stabilizing elements content and decrease with increasing $K_β$ value.

For the intermediate cooling rates, down to 3.5°C s^{-1}, martensitic transformation is accompanied by diffusional transformation β→α + β and the volume fraction of martensitic phases decreases to the benefit of stable α and β phases (Figs 1-2). Cooling rates below 2°C s^{-1} lead to a diffusion controlled nucleation and growth of stable α and β phases in the shape of colonies of parallel α-phase lamellae in primary β-phase grains (Fig. 5). For extremely low cooling rates precipitations of $TiCr_2$ phase were identified in the Ti-6Al-2Mo-2Cr alloy which were formed in eutectoid transformation.

In the transition alloy Ti-6Al-5Mo-5V-1Cr-1Fe martensitic transformation was not observed at any cooling rate. High cooling rate (>18°C s^{-1}) results in metastable $β_M$ microstructure. At lower cooling rates α-phase precipitates as a result of diffusional transformation. At lowest cooling rate, similarly to Ti-6Al-2Mo-2Cr alloy eutectoid transformation occurs and traces of $TiCr_2$ and $TiFe_2$ appears [7].

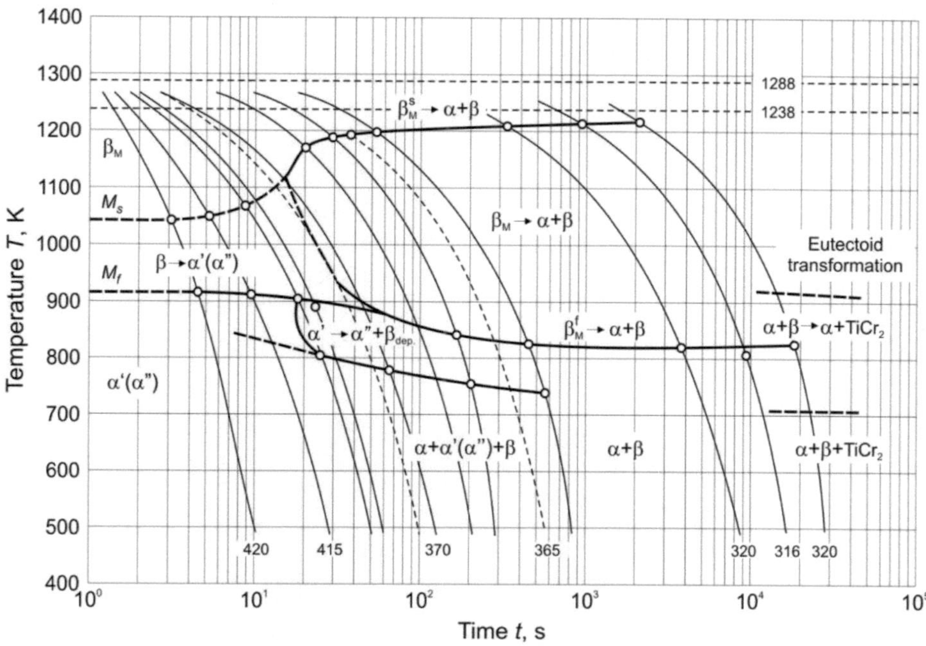

Figure 2. CCT diagram for Ti-6Al-2Mo-2Cr alloy.

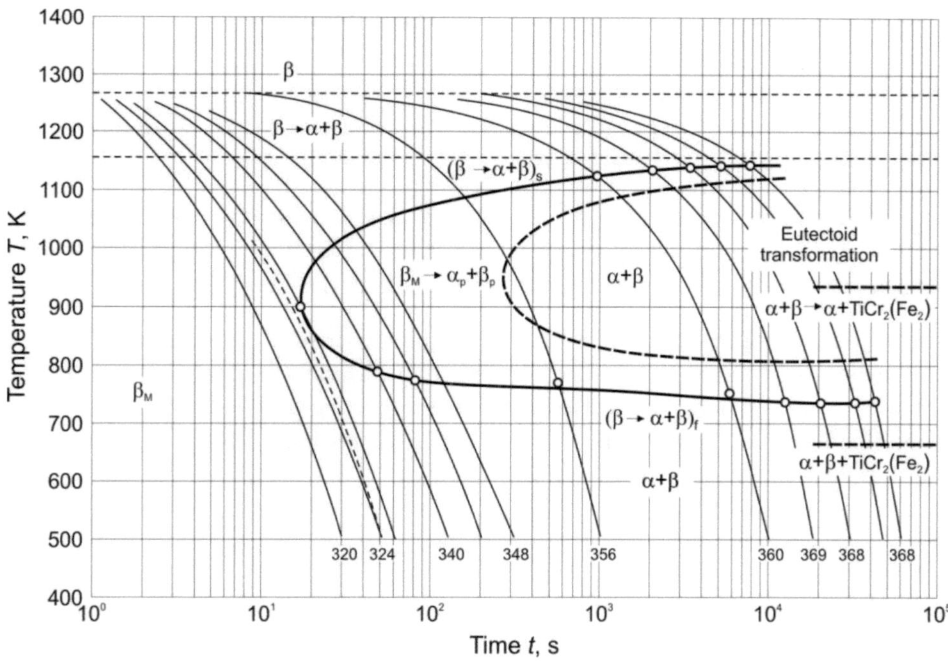

Figure 3. Fig. 3.CCT diagram for Ti-6Al-5Mo-5V-1Cr-1Fe alloy.

Cooling rate, °C·s^{-1}	Phase composition of the alloy		
	Ti-6Al-4V	Ti-6Al-2Mo-2Cr	Ti-6Al-5Mo-5V-1Cr-1Fe
48-18	α'(α")	α'(α")	$β_M$
9	α + α'(α")	α + α'(α") + β	$β_M$ + α
7	α + α'(α")	α + α'(α")$_{trace}$ + β	$β_M$ + α
3.5	α + α'(α")$_{trace}$ + β	α + α'(α")$_{trace}$ + β	$β_M$ + α
1.2-0.04	α + β	α + β	α + β
0.024-0.004	α + β	α + β + $TiCr_2$	α + β + $TiCr_2(Fe_2)$

Table 3. Phase composition of the selected titanium alloys after controlled cooling from the β-phase range [6,9]

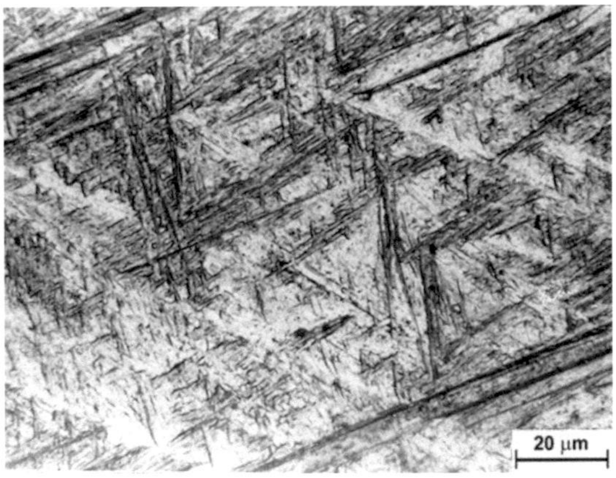

Figure 4. Martensitic microstructure of Ti-6Al-2Mo-2Cr alloy after cooling from 1050°C at a rate of 48°C s^{-1} (LM–DIC micrograph).

The important parameters for a lamellar microstructure with respect to mechanical properties of the alloy are the β-grain size, size of the colonies of α-phase lamellae, thickness of the α-lamellae and the morphology of the interlamellar interface (β-phase) (Fig. 6) [12-13].

Increase in cooling rate leads to refinement of the microstructure – both α colony size and α-lamellae thickness are reduced. Additionally new colonies tend to nucleate not only on β-phase boundaries but also on boundaries of other colonies, growing perpendicularly to the existing lamellae. This leads to formation of characteristic microstructure called "basket weave" or Widmanstätten microstructure (Fig. 7) [3].

4. Tensile and fatigue properties

Mechanical properties of two phase titanium alloys strongly depend on morphology of particular phases. In the case of the alloys with lamellar microstructure, the thickness of α lamellae and diameter of their colonies have the most significant influence [3,14].

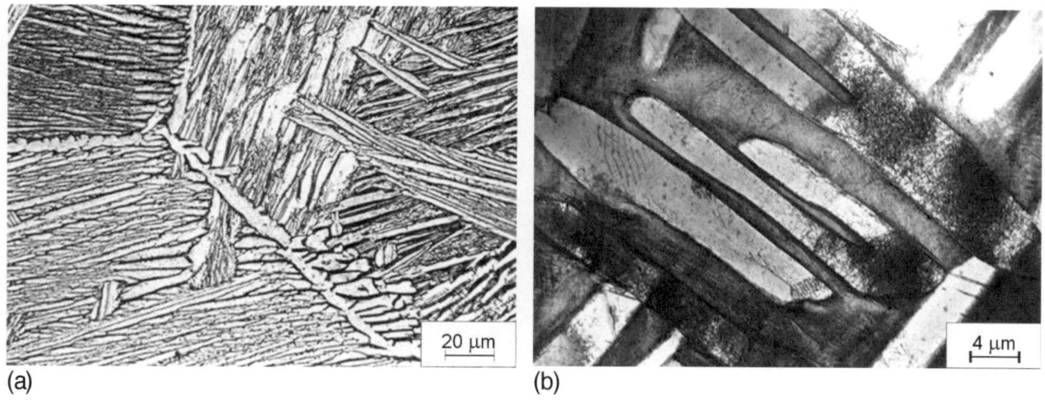

Figure 5. Microstructure of Ti-6Al-2Mo-2Cr alloys after cooling from 1050°C at a rate of 1.2°C s^{-1}: a) LM micrograph, b) TEM micrograph.

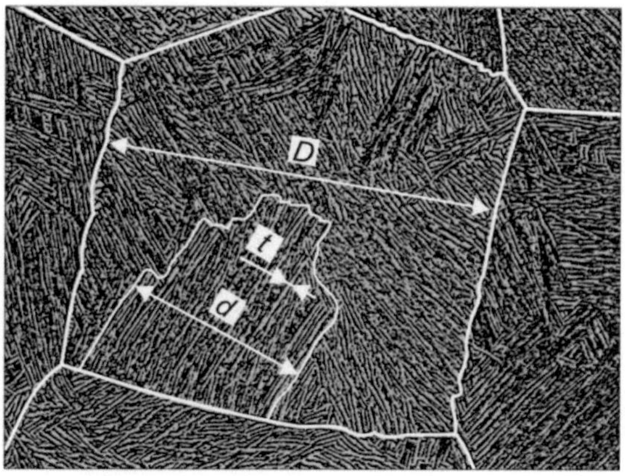

Figure 6. Stereological parameters of lamellar microstructure: *D* – primary β-phase grain size, *d* – size of the colony of parallel α-lamellae, *t* – thickness of α-lamellae.

Refinement of the microstructure results in higher yield stress (Fig. 8a). However the increase of yield stress is moderate unless martensitic phase is present. Tensile elongation increases with increasing cooling rate at first (Fig. 8b). However, after reaching maximum the ductility curve declines. Such behaviour was reported earlier and attributed to the change of fracture mode from ductile transcrystalline for low cooling rates to ductile intercrystalline fracture along continuous α phase layers at primary β grain boundaries [6,8].

The size of the colonies of α lamellae having the same crystallographic orientation have significant influence on the mechanical properties of the alloy as it is a measure of effective slip length [8,15]. However transition to the 'basket weave' type of microstructure makes the determination of colonies size even more difficult. Because of that the thickness of α-lamellae was also taken into account as the quantitative parameter illustrating the effect of microstructure refinement on mechanical properties.

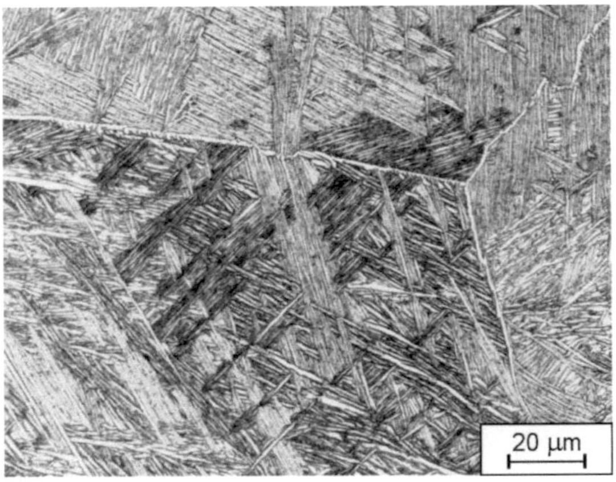

Figure 7. "Basket-weave" or Widmanstätten microstructure of Ti-6Al-4V alloy after cooling from β-phase range at the rate of 9°C s^{-1}.

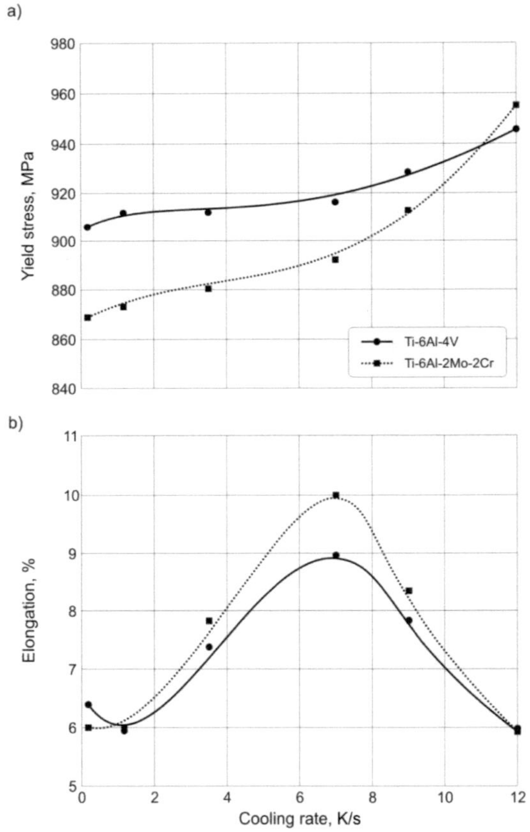

Figure 8. Yield stress and tensile elongation dependence on the cooling rate from β-phase range for selected titanium alloys.

Average thickness of α-phase lamellae, μm	YS, MPa	UTS, MPa	σ_f, MPa	HV
2.4	970	1115	565	326
3.0	928	1068	580	330
5.5	916	1056	570	325
7.6	908	1038	560	336

σ_f – fatigue strength at 10^7 cycles in rotational bending test.

Table 4. Mechanical properties of the Ti-6Al-4V alloy

Average thickness of α-phase lamellae, μm	YS, MPa	UTS, MPa	σ_f, MPa	HV
1.7	980	1136	550	340
2.0	944	1108	575	338
3.6	924	1055	560	336
6.2	922	1024	540	332

Table 5. Mechanical properties of the Ti-6Al-2Mo-2Cr alloy.

Average thickness of α-phase lamellae, μm	YS, MPa	UTS, MPa	σ_f, MPa	HV
0.8	1235	1305	540	348
1.5	1225	1285	555	340
2.6	1186	1262	535	342
4.3	1160	1236	520	336

Table 6. Mechanical properties of the Ti-6Al-5Mo-5V-1Cr-1Fe alloy.

Following values of geometrical parameters of lamellar α-phase, i.e. thickness of the α-lamellae (t) and diameter of the α-phase lamellae colony (d), provided maximum fatigue strength of the investigated alloys:

- Ti-6Al-2Mo-2Cr $t = 2$ μm, $d = 20$ μm,
- Ti-6Al-4V $t = 3$ μm, $d = 30$ μm,
- Ti-6Al-5Mo-5V-1Cr-1Fe $t = 1.5$ μm, $d = 35$ μm.

Fatigue fracture surfaces showed transgranular character with typical ductile surroundings of β-phase around α-phase (Fig. 9). Size of the dimples were closely related to thickness of the α-lamellae and size of the colonies of parallel α-lamellae [16]. No pronounced beach markings or striations were identified which is an evidence of frequent change of the crack growth direction. This phenomenon along with secondary crack branching are important reasons for advantageous effect of lamellar microstructure on fatigue behaviour.

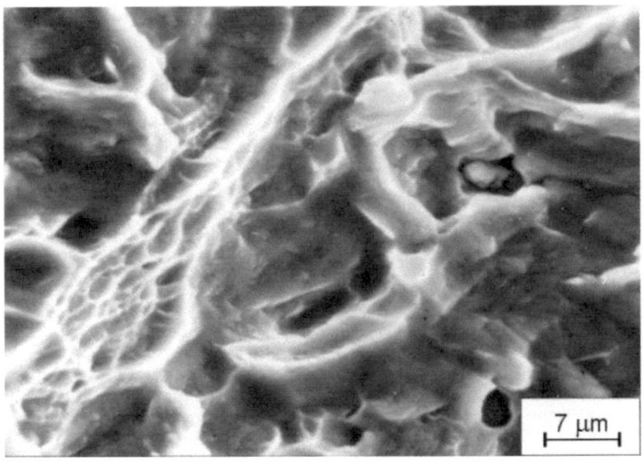

Figure 9. Fatigue fracture surfaces of Ti-6Al-5Mo-5V-1Cr-1Fe alloy cooled from 1020°C at a rate of 0.8°C s^{-1}.

The progress of the crack tip through regions of interfacial β-phase is accompanied by the absorption of large amount of energy due to intensive plastic deformation, contributing to lowering the rate of crack propagation. When thickness of β-phase regions decreases, it cannot absorb sufficient amounts of energy and retard the crack propagation.

5. Summary

Microstructure of two-phase titanium alloys after deformation or heat treatment carried out at a temperature in the range of β-phase stability depends on cooling rate. High cooling rates (>18°C s^{-1}) result in martensitic $\alpha'(\alpha'')$ microstructure for alloys having β stability factor $K_\beta<1$ and metastable β_M microstructure for alloys with higher contents of β-stabilizers. Low and moderate cooling rates lead to development of lamellar microstructure consisting of colonies of α-phase lamellae within large β-phase grains. Decrease of cooling rate cause increase both in thickness of individual α-phase lamellae and size of the colonies of parallel α-lamellae. This in turn lowers yield stress and tensile strength of the alloys.

Lamellar α-phase microstructure of the alloy heat treated in the β-range has beneficial effect on its fatigue behaviour. This is the result of frequent change in crack direction and secondary crack branching. When α-phase lamellae are too large thin layers of β-phase are not capable to absorb large amounts of energy and retard the crack propagation. In this case the colony of the α-phase lamellae behaves as singular element of the microstructure. This phenomenon is more intensive in the alloy with smaller value of K_β coefficient (Ti-6Al-4V). Sufficient thickness of β-phase surroundings enables absorption of energy in the process of plastic deformation of regions ahead of the crack tip, contributing to slowing the rate of crack propagation and therefore increasing fatigue life.

Author details

J. Sieniawski, W. Ziaja, K. Kubiak and M. Motyka

Rzeszów University of Technology, Dept. of Materials Science, Poland

References

[1] Bylica A, Sieniawski J. Titanium and Its Alloys. PWN, Warsaw, Poland, (1985). (in Polish).

[2] Williams J. C, Starke Jr. E. A. Progress in structural materials for aerospace systems. Acta Materialia. (2003). 51(19):5775-5799.

[3] Lutjering G, Williams J C. Titanium. Springer-Verlag, Berlin, 2007.

[4] Lütjering G. Property optimization through microstructural control in titanium and aluminum alloys. Materials Science and Engineering. (1999). A263(1-2):117–126.

[5] Markovskya P. E, Semiatin S. L. Tailoring of microstructure and mechanical properties of Ti–6Al–4V with local rapid (induction) heat treatment. Materials Science and Engineering. (2011). A528(7-8):3079–3089.

[6] Sieniawski J, Filip R, Ziaja W. The effect of microstructure on the mechanical properties of two phase titanium alloys. Materials & Design. (1997). 18(4-6):361-363.

[7] J. Sieniawski, Phase transformations and microstructure development in multicomponent titanium alloys containing Al, Mo, V and Cr, Oficyna Wydawnicza Politechniki Rzeszowskiej, Rzeszów, (1985). (in Polish).

[8] Lütjering G. Influence of processing on microstructure and mechanical properties of (+) titanium alloys. Materials Science and Engineering. (1998). A243(1-2):32-45.

[9] Kubiak K, Sieniawski J. Development of the microstructure and fatigue strength of two-phase titanium alloys in the processes of forging and heat treatment. Journal of Materials Processing Technology. (1998) 78(1-3):117-121.

[10] Nalla R. K, Boyce B. L, Campbell J. P, Peters J. O, Ritchie R. O. Influence of microstructure on high-cycle fatigue of Ti-6Al-4V: bimodal vs. lamellar structures. Metallurgical and Materials Transactions. (2002). 33A(3):899-918.

[11] Moiseyev VN. Titanium alloys. Russian aircraft and aerospace applications. Taylor & Francis. New York. (2006).Tiley J, Searles T, Lee E, Kar S, Banerjee R, Russ J. C, Fraser H. L. Quantification of microstructural features in / titanium alloys. Materials Science and Engineering. (2004). A372(1-2):191-198.

[12] Filip R, Kubiak K, Ziaja W, Sieniawski J. The effect of microstructure on the mechanical properties of two-phase titanium alloys, Journal of Materials Processing Technology. (2003). 133(1-2):84-89.

[13] Sieniawski J, Grosman F, Filip R, Ziaja W. Microstructure factors in fatigue damage process of two-phase titanium alloys. Titanium '95 Science and Technology, P.A. Blenkinsop, W.J. Evans and H.M. Flower eds., The Institute of Materials, Birmingham. (1996). 1411-1418.

[14] Gil F. J, Manero J. M, Ginebra M. P, Planell J. A. The effect of cooling rate on the cyclic deformation of -annealed Ti-6Al-4V. Materials Science and Engineering. (2003). A349(1-2):150-155.

[15] Ziaja W, Sieniawski J, Kubiak K, Motyka M.: Fatigue and microstructure of two phase titanium alloys. Inżynieria Materiałowa. (2001). 22(3):981-985.

Chapter 5

Formability Characterization of Titanium Alloy Sheets

F. Djavanroodi and M. Janbakhsh

Additional information is available at the end of the chapter

http://dx.doi.org/10.5772/55889

1. Introduction

Recently, different industries faced the challenge of implementation of titanium alloys in order to produce components with different formability characteristics. Titanium alloy sheets are defined as hard-to-form materials regarding to their strength and formability characteristics. Consequently, in order to soundly manufacture a part made from the mentioned alloys, novel processes such as hydroforming, rubber pad forming and viscous pressure forming instead of conventional stamping or deep drawing are applied.

In sheet metal forming industries, FE simulations are commonly used for the process/tool design. Availability of suitable mechanical properties of the sheet material is important factor for obtaining accurate FE simulations results.

1.1. Biaxial bulge test

The most commonly used method to investigate the flow stress curve is uniaxial tensile test in which true stress-true strain curve is expressed in uniaxial stress state. However, maximum plastic strains obtained in uniaxial loading condition is not sufficient for most sheet metal forming simulation processes which involve biaxial state of stress [1-5]. Hydraulic bulge test is a comparative test method in which biaxial stress-strain curve could be attained. In 1950, a key theoretical pillar for the hydraulic bulge test was established by Hill [6]. In his study, Hill assumed a circular profile for the deforming work piece which allowed for the introduction of a closed form expression for the thickness at the pole region [5].

The experimental bulge test method involves pumping hydraulic fluid [4, 7-10] or a viscous material as a pressure medium [2-3, 11] instead of a hydraulic fluid into the die cavity. Circular as well as elliptical dies can also be used to determine anisotropy coefficients of material in different directions with respect to rolling direction [12-15].

1.2. Forming Limit Diagrams (FLD)

As mentioned before, due to increased demand for light weight components in aerospace, automotive and marine industries, recently, titanium sheet alloys have gained ever more interests in production of structural parts. In order to better understand the cold formability of these alloys, their behavior in sheet metal forming operations must be determined both experimentally and theoretically. Sheet metal formability is often evaluated by forming limit diagram. The concept of FLD was first introduced by Keeler [16] and Goodwin [17]. Forming limit diagram provides the limiting strains a sheet metal can sustain whilst being formed. Laboratory testing has shown that the forming limit diagrams are influenced by several factors including strain hardening exponent and anisotropy coefficients [18-20], strain rate [21-23], temperature [24], grain size and microstructure [25-26], sheet thickness [27], strain path changes [28-29], and heat treatment [30].

In recent years many experimental techniques have been developed to investigate the FLDs from different aspects [31-34]. These studies were based on elimination of frictional effects resulted from toolsets and materials, the uniformity of the blank surface and mechanical properties of sheet materials deduced from the conventional tensile testing.

The available tests for the determination of FLDs include: hydraulic bulge test [35], Keeler punch stretching test [36], Marciniak test [37], Nakazima test [38], Hasek test [39] and the bi-axial tensile test using cruciform specimen [40] (in short cruciform testing device). From previous studies [36-39], it is widely acknowledged that friction remains an unknown factor yet to be effectively characterized and understood. Thus, the list of available tests is greatly reduced to only two options - hydraulic bulge test and cruciform testing device. Further analysis shows that due to simplicity of equipment and specimen (i.e. less costly), hydraulic bulge test is comparatively preferred [41].

On the other hand, several researchers have proposed a number of analytical models to predict FLD. Hill's localized instability criterion [42], combined with Swift's diffused instability criterion [43] was the first analytical approach to predict FLDs. It was shown that forming limit curves are influenced by material work hardening exponent and anisotropy coefficient. Xu and Wienmann [44] showed that for prediction of the FLD, the shape of the used yield surface had a direct influence on the limit strains. They used the Hill'93 criterion to study the effect of material properties on the FLDs. The M-K model [45] predicts the FLD based on the assumption of an initial defect in perpendicular direction with respect to loading direction. The assumption made for this non-homogeneity factor is subjective and hence, the forming limit diagram is directly influenced by it. This method was then developed considering the material properties [46-47].

Several researchers used the ductile fracture criteria for forming limit predictions in hydro-forming process [48], deep drawing process [49], bore-expanding [50] and biaxial stretching [51]. Fahrettin and Daeyong [52] and Kumar et al. [53] proposed the thickness variations and the thickness gradient criterion respectively. These criteria are limited because they require precise measurements of thickness. Bressan and Williams [54] used the method of shear instability to predict the FLDs. Consequently, experimental techniques are widely accepted for determination of the FLDs and for verification of the predicted FLDs resulted from analytical models.

Djavanroodi and Derogar [19] used the Hill-Swift model to predict the limit strains for titanium and aluminum sheets. They performed hydroforming deep drawing test in conjunction with a novel technique called "floating disc" to determine the FLDs experimentally. It was concluded from their work that as strain hardening exponent and anisotropic coefficients increase, the limit strains will also increase, and consequently, this allows the FLD to be shifted up.

In this chapter, different analytical approaches as well as the experimental methods are applied to obtain the uniaxial and biaxial flow stress curves for Ti-6Al-4V sheet metal alloy. The hydraulic bulge test was carried out and findings were compared with the results obtained from the uniaxial tensile test. Circular-shaped die was used. Stepwise test with gridded specimens and continuous experiments were performed. For flow stress calculations, both the dome height and pressure were measured during the bulge tests. The effects of anisotropy and strain hardening on material formability were also investigated.

On the other hand, a practical approach was implemented for experimental determination of FLD and several theoretical models for prediction of forming limit diagrams for 1.08mm thick Ti-6Al-4V titanium sheet alloy subjected to linear strain paths were applied. For the experimental approach, the following test pieces have been used to obtain different regions of FLD: circle specimens to simulate biaxial stretching region of FLD (positive range of minor strain); non-grooved tensile specimens (dog-bone shaped specimens) to simulate the uniaxial strain path and two distinctive grooved tensile specimens representing the strain path ranging from uniaxial tension to plane strain region of FLD. The onset of localized necking was distinguished by investigating the strain distribution profiles near the necking region. Furthermore to predict the theoretical FLDs, Swift model with Hill93 yield criteria [55] and M-K model with Hill93 and BBC2000 yield criteria [56] were used. Predicted FLDs were compared with the experimental data to evaluate the suitability of the approaches used. Moreover, considering the extensive application of *Autoform 4.4* software in sheet metal forming industries, several parts representing different strain paths were formed to evaluate the FLDs of the tested sheets numerically. The effects of process parameters as well as yield loci and material properties used in simulations were also discussed.

2. Theoretical approaches

2.1. Hydroforming bulge test

Hydroforming bulge test is one of the most commonly used balanced biaxial tests in which a circular sheet metal fully clamped between two die surfaces is drawn within a die cavity by applying hydrostatic pressure on the inner surface of the sheet. The die cavity diameter ($d_d = 2R_d$), the upper die fillet radius (R_f) and initial sheet thickness (t_0) are constant parameters of any hydroforming bulge testing. Instantaneous variables of biaxial test are: bulge pressure (p), dome height (h_b), bulge radius (R_b) and sheet thickness at the dome apex (t). The schematic view of hydroforming bulge test is shown in Fig.1.

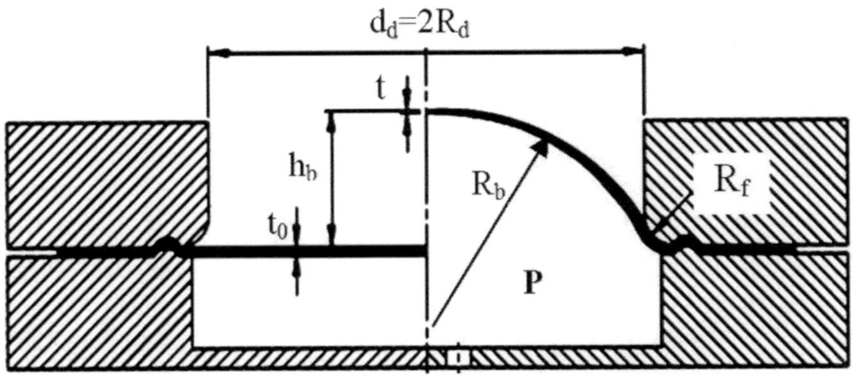

Figure 1. Scheme of the hydroforming bulge test

In order to obtain flow stress curve, first, a combination of constant and variable parameters are introduced to several equations proposed by the other researchers to calculate the instantaneous bulge radius [6, 57] and the sheet thickness [6, 58-59] at the dome apex. Subsequently, by making the assumption that during the bulging process the sheet metal behavior is the same as thin-walled structure and by implementing the classical membrane theory, the flow stress curves are obtained. Eqs.1 and 2 represent the theories for calculating the bulge radius proposed by Hill [6] and Panknin [57], respectively:

$$R_b = \frac{d_d + 4h_b}{8h_b} \tag{1}$$

$$R_b = \frac{\left(R_d + R_f\right)^2 + h_b^2 - 2R_f h_b}{2h_b} \tag{2}$$

Eqs.3 to 5 represents the theories for calculating the instantaneous sheet thickness at the dome apex proposed by Hill [6], Chakrabarty et al. [58] and Kruglov et al. [59], respectively. In this chapter, Eqs.1 to 5 was used in the theoretical approach.

$$t = t_0 \left(\frac{1}{1 + \left(h_b / R_d\right)^2} \right)^2 \tag{3}$$

$$t = t_0 \left(\frac{1}{1 + \left(h_b / R_d\right)^2} \right)^{2-n} \tag{4}$$

http://dx.doi.org/10.5772/55889

$$t = t_0 \left[\frac{(R_d / R_b)}{Sin^{-1}(R_d / R_b)} \right]^2 \tag{5}$$

The second stage for calculating the flow stress curves for sheet material is implementing the classical membrane theory. Due to very low ratio of thickness to radius of the sheet (t/R_d<<0.1), the stress component in perpendicular direction to sheet surface is not considered (σ_z=0). By considering Tresca's yield criterion, Gutscher et al. [2], proposed an equation to evaluate the effective stress resulted from the hydroforming bulge test:

$$\bar{\sigma}_{isotropic} = \frac{p}{2}(\frac{R_b}{t} + 1) \tag{6}$$

Principle strains at dome are: ε_θ, ε_ϕ and ε_t. Assuming Von-Mises yield criterion and equality $\varepsilon_\theta = \varepsilon_\phi$, the effective strain can be calculated as:

$$\bar{\varepsilon}_{isotropic} = \sqrt{\frac{2}{9}\left[(\varepsilon_\theta - \varepsilon_\varphi)^2 + (\varepsilon_\theta - \varepsilon_t)^2 + (\varepsilon_\varphi - \varepsilon_t)^2 \right]} \tag{7}$$

It is known that due to the principle of volume constancy ($\varepsilon_\theta + \varepsilon_\phi + \varepsilon_t = 0$), plastic deformation does not yield any volume change [2, 60]:

$$\bar{\varepsilon}_{isotropic} = \varepsilon_\theta + \varepsilon_\phi = -\varepsilon_t = \ln\frac{t_0}{t} \tag{8}$$

Since due to rolling conditions, sheet metal properties differ in various directions with respect to rolling direction (anisotropy), effective stress and effective strain components should be corrected for anisotropy. In Equations 6 and 8, no anisotropy correction was introduced. Consequently, assuming Hill'48 yield criterion in conjunction with plane stress assumption, Smith et al. [5], proposed equations 9 and 10 for determination of effective stress and effective strains for sheet metals considering the average normal anisotropy (*R*), respectively.

$$\bar{\sigma}_{anisotropic} = \bar{\sigma}_{isotropic} \left[2 - \frac{2R}{(R+1)} \right]^{1/2} \tag{9}$$

$$\bar{\varepsilon}_{anisotropic} = \frac{2\varepsilon_{isotropic}}{(2 - (2R / R + 1))^{1/2}} \tag{10}$$

2.2. Forming Limit Diagrams

Hill93 and BBC2000 constitutive models

In this paper Hill 93 and BBC2000 constitutive models were used to predict the FLDs. Eq.11 represents Hill'93 yield criterion [55]:

$$\frac{\sigma_1^2}{\sigma_0^2} - \frac{c\sigma_1\sigma_2}{\sigma_0\sigma_{90}} + \frac{\sigma_2^2}{\sigma_{90}^2} + \left[(p+q) - \frac{(p\sigma_1 + q\sigma_2)}{\sigma_b}\right]\frac{\sigma_1\sigma_2}{\sigma_0\sigma_{90}} = 1 \tag{11}$$

$$\frac{c}{\sigma_0\sigma_{90}} = \frac{1}{\sigma_0^2} + \frac{1}{\sigma_{90}^2} - \frac{1}{\sigma_b^2} \qquad \begin{aligned} p &= \left[\frac{2R_0(\sigma_b - \sigma_{90})}{(1+R_0)\sigma_0^2} - \frac{2R_{90}\sigma_b}{(1+R_{90})\sigma_{90}^2} + \frac{c}{\sigma_0}\right]\frac{1}{\frac{1}{\sigma_0} + \frac{1}{\sigma_{90}} - \frac{1}{\sigma_b}} \\ q &= \left[\frac{2R_{90}(\sigma_b - \sigma_0)}{(1+R_{90})\sigma_{90}^2} - \frac{2R_0\sigma_b}{(1+R_0)\sigma_0^2} + \frac{c}{\sigma_{90}}\right]\frac{1}{\frac{1}{\sigma_0} + \frac{1}{\sigma_{90}} - \frac{1}{\sigma_b}} \end{aligned} \tag{12}$$

Where c, p and q are Hill'93 coefficients and can be calculated using five mechanical parameters obtained from two uni-axial tensile tests and an equi-biaxial tension (Eq.12).

Banabic et. al [56] proposed a new yield criterion called BBC2000 for orthotropic sheet metals under plane stress conditions. The equivalent stress is defined as:

$$\bar{\sigma} = \left[a(b\Gamma + c\Psi)^{2k} + a(b\Gamma - c\Psi)^{2k} + (1-a)(2c\Psi)^{2k}\right]^{\frac{1}{2k}} \tag{13}$$

Where a, b, c and k are material parameters, while Γ and Ψ are functions of the second and third invariants of a fictitious deviatoric stress tensor and can be expressed as explicit dependencies of the actual stress components:

$$\begin{aligned} \Gamma &= (d+e)\sigma_{11} + (e+f)\sigma_{22} \\ \Psi &= \sqrt{\left[\frac{1}{2}(d-e)\sigma_{11} + \frac{1}{2}(e-f)\sigma_{22}\right]^2 + g^2\sigma_{xy}^2} \end{aligned} \tag{14}$$

Where d,e, f and g are anisotropy coefficients of material and k-value is set in accordance with the crystallographic structure of the material (k=3 for BCC alloys and k=4 for FCC alloys). For the BBC2000 yield model, the detailed description can be found in [56].

Theoretical prediction of the FLD

The simulation of plastic instability is performed using M–K and Hill-Swift analysis. The rigid plastic material model with isotropic work hardening and the plane stress condition were assumed.

A detailed description of the theoretical M–K analysis, schematically illustrated in Fig.2, can be found in [45]. The M-K model is based on the growth of an initial defect in the form of a narrowband perpendicular to the principal axis. The initial value of the geometrical defect (f_0) is characterized by Eq.15, where $t^a{}_0$ and $t^b{}_0$ are the initial thicknesses in the homogeneous and grooved region, respectively.

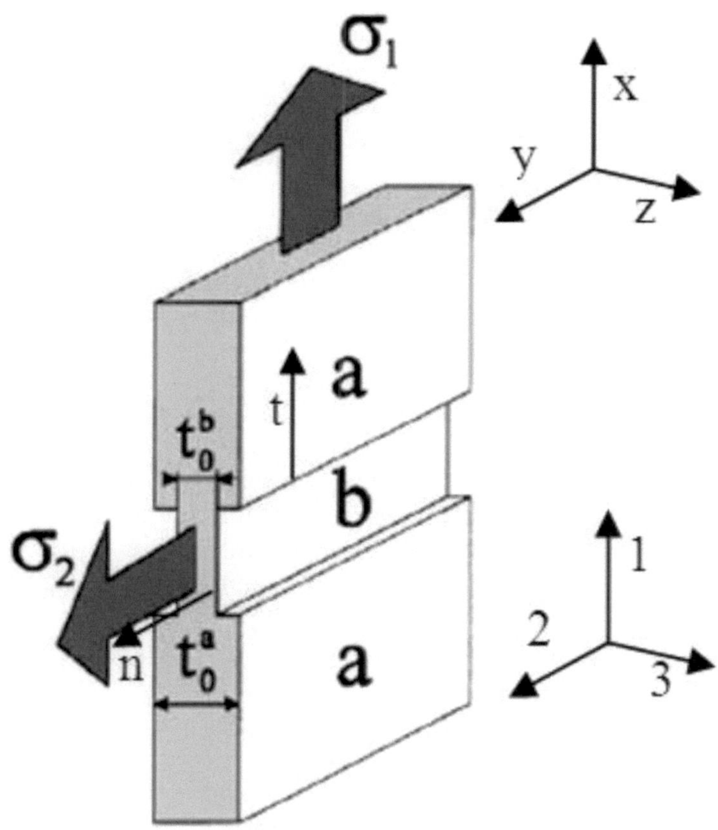

Figure 2. Schematic description of M-K model

$$f_0 = \frac{t_0^b}{t_0^a} \tag{15}$$

The x, y, z-axes correspond to rolling, transverse and normal directions of the sheet, whereas 1 and 2 represent the principal stress and strain directions in the homogeneous region. The set of axis bound to the groove is represented by *n, t, z*-axes where 't' is the longitudinal one. The plastic flow occurs in both regions, but the evolution of strain rates is different in the two zones. When the flow localization occurs in the groove at a critical strain in homogeneous region, the limiting strain of the sheet is reached. Furthermore, the major strain is assumed to occur along the X-axis. M–K necking criterion assumes that the plastic flow localization occurs when the equivalent strain increment in imperfect region (b) reaches the value ten times greater than in homogeneous zone (a) ($d\varepsilon_b > 10 d\varepsilon_a$). When the necking criterion is reached the computation is stopped and the corresponding strains (ε_a^{xx}, ε_a^{yy}) obtained at that moment in the homogeneous zone are the limit strains. For the model, equation expressing the equilibrium of the forces acting along the interface of the two regions could be expressed as follow:

$$\sigma_{1a} t_a = \sigma_{1b} t_b \tag{16}$$

Strains parallel to the notch are equal in both regions:

$$d\varepsilon_{2a} = d\varepsilon_{2b} \tag{17}$$

In addition, the model assumes that the strain ratio in zone *a*, is constant during the whole process:

$$d\varepsilon_{2a} = \rho d\varepsilon_{1a} \tag{18}$$

Detailed FLD calculation using BBC2000 yield criterion is presented in [61]. Swift work-hardening model with strain rate sensitivity factor was used:

$$\bar{\sigma} = K\left(\varepsilon_0 + \bar{\varepsilon}\right)^n \dot{\varepsilon}^m \tag{19}$$

In this chapter, in addition to M-K analysis, the Swift analysis [43] with Hill'93 yield criterion was used to predict the FLDs [62-63].

3. Experimental work

3.1. Tensile test

Tensile test specimens were cut according to ASTM E8 standard. At least two samples at each direction (0°,45°,90°) with respect to rolling directions were tested according to ASTM E517-00

http://dx.doi.org/10.5772/55889

standard [64]. Tensile test was carried out under constant strain rate of 1×10^{-3} s^{-1} at room temperature.

Although r-value is introduced as the ratio of width strain ε_w to thickness strain ε_t, the thickness strain in thin sheets can not be accurately measured. Hence, by measuring longitudinal ε_l and width strains and also by implementing the principle of volume constancy (Eq.20), the thickness strain was obtained as follows (Eq.21):

$$\varepsilon_l + \varepsilon_w + \varepsilon_t = 0 \tag{20}$$

$$\varepsilon_t = -\left(\varepsilon_l + \varepsilon_w\right) \tag{21}$$

The strain ratio (r-value) was calculated for all the materials at different direction (0, 45 and 90° to the rolling direction) (Eq.22). Subsequent to that, normal anisotropy R (Eq.23) as well as planar anisotropy ΔR (Eq.24) were calculated according to ASTM E517-00 [64].

$$R_x = \frac{\varepsilon_{w,x}}{\varepsilon_{t,x}} \tag{22}$$

$$R = \frac{R_0 + 2R_{45} + R_{90}}{4} \tag{23}$$

$$\Delta R = \frac{R_0 + R_{90} - 2R_{45}}{2} \tag{24}$$

Where: x is the angle relative to the rolling direction (0°, 45°, 90°).

3.2. Hydroforming bulge test

The experimental apparatus used to conduct the hydraulic bulge test is composed of a tooling set, a hydraulic power generator and measurement devices. For the assembled toolset, maximum forming pressure can reach 500bars. To avoid any oil leakage during the forming process, a rubber diaphragm was placed between the conical part of the die and the conjunctive disc. A pressure gage and a dial indicator were used to measure the chamber pressure and bulge height, respectively during the bulging process. The indicator used in the experiments was delicate and could not withstand impact loads as the specimen bursts. Hence, for bulge testing of titanium sheets, at least three samples were burst in the absence of the indicator to discern the bursting pressures. Other samples were tested up to 90-95% of bursting pressure while the indicator was used to measure the bulge height during the process. In order to ensure pure stretching, the pre-fabricated draw bead was implemented at the flange area of the bulge samples. Consequently, pure stretching of the sheet material was obtained during the bulging process. Measuring devices were also calibrated before the test to ensure precise measurements. For bulge testing of sheet materials a die set was used. For bulge testing of Ti-6Al-4V, a large die set was designed and manufactured in order to reach the bursting pressure through available hydraulic pressure unit. Table 1 shows dimensions of the die set in addition to specifications of hydroforming bulge test apparatus. Hydroforming die used for bulge testing is shown in Fig.3.

	Die set specification
Bulge diameter ($2R_d$)	90 mm (3.54 in)
Die fillet radius	6 mm (0.236 in)
Maximum chamber pressure	50 MPa (7400 psi)
Flow rate	2.5 lit/min

Table 1. Specifications of hydroforming bulge apparatus.

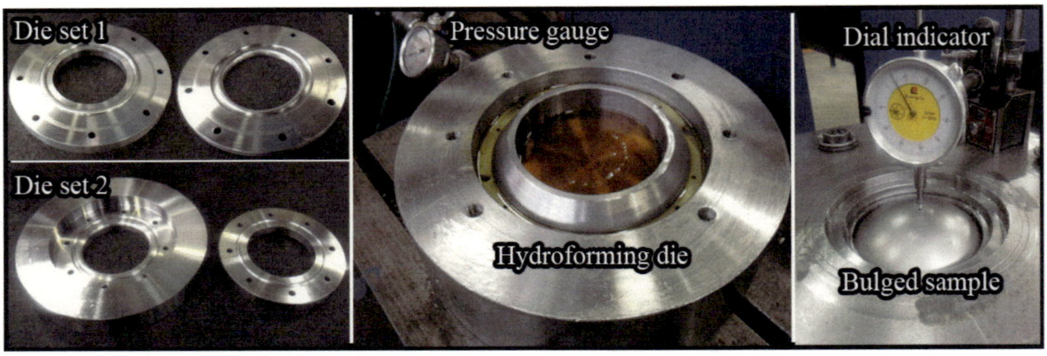

Figure 3. Hydroforming bulge test apparatus

3.3. Forming limit diagrams

In order to evaluate the FLDs, different strain paths, which cover full domain of the FLD, were examined and shown in Fig.4. These paths are spanned between the uniaxial tension region (ε_1=-2ε_2) and the equi-biaxial stretching (ε_1=ε_2). The linear strain paths are described through the strain ratio (Eq.25) parameter representative of the strain state.

$$\rho = \frac{d\varepsilon_2}{d\varepsilon_1} \tag{25}$$

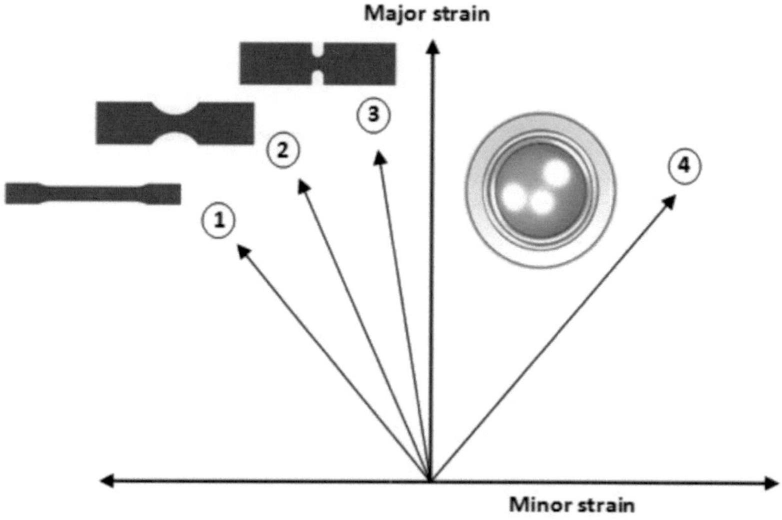

Figure 4. Different types of specimen representative of the linear strain paths

3.4. Preparation of the specimens

Three different shapes of tensile specimens were prepared using wire EDM. According to Fig. 4, a non-grooved specimen was prepared to simulate the strain path#1, two distinctive grooved specimens were used to obtain draw points representing the strain paths#2 and 3. The non-grooved and grooved specimens were then drawn under a linear load. The specimens are shown in Figures 5-7 and the dimensions are listed in Table 2.

In order to obtain the tension-tension side of the FLD (pure stretching region), the bulge test specimens were prepared. The bulge samples for simulation of the strain path#4, were 160mm in diameter and the cavity diameter of circular dies was 90mm (as shown in Table 1). LASER imprinting technique was used to print a grid of circles on the surface of the tension and the bulging samples. The circles were 2mm in diameter. The diameter of the circles of the grids have been measured before and after the deformation throughout the major and minor principal directions taking as reference a perpendicular axis system placed in the geometric centre of each circle or in the centre of the tension specimen. These principal directions are parallel and perpendicular, respectively, to the rolling direction of the sheet.

Tensile specimens were tested up to the fracture point. Likewise, bulge testing of the bulge samples were carried out to reach the bursting point. Subsequent to that, diameters of ellipses which were the conclusions of deformed circles were measured precisely.

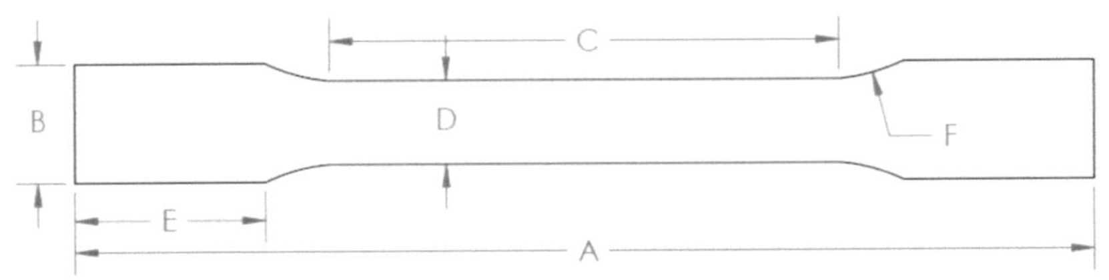

Figure 5. Uniaxial tensile specimen representative of strain path#1 (no.1).

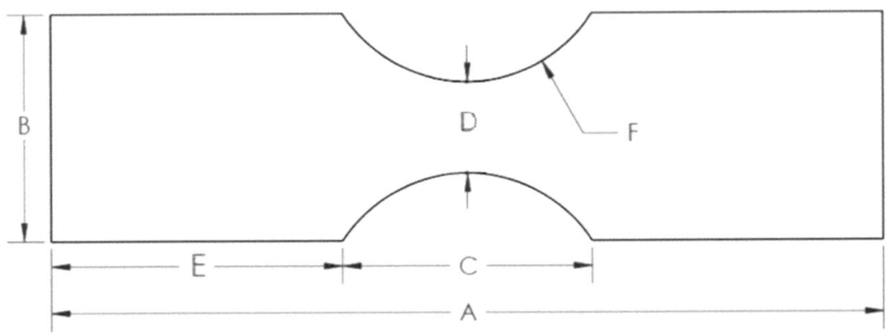

Figure 6. Grooved specimen representative of strain path#2 (no.2).

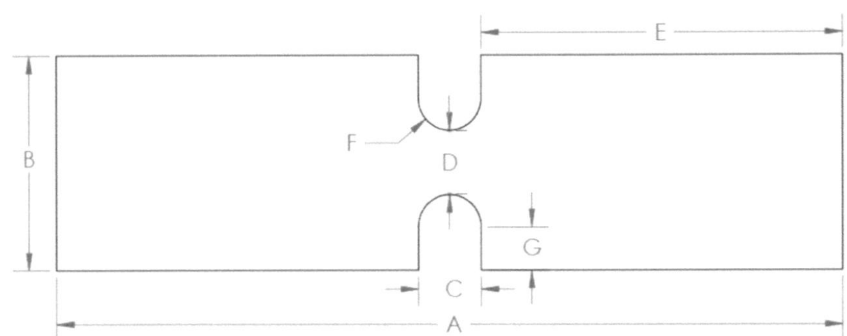

Figure 7. Grooved specimen representative of strain path#3 (no.3).

Specimens no.	A (mm)	B (mm)	C (mm)	D (mm)	E (mm)	F (mm)	G (mm)
1	150	17	75	12	28	R20	-
2	150	40	45	16	52.5	R26	-
3	150	40	12	12	69	R6	8

Table 2. Dimensions of the tensile specimens

http://dx.doi.org/10.5772/55889

4. Numerical approach

More recently, several researchers [19, 65-67] have investigated the forming limit diagrams through finite element codes. In this chapter, *Autoform Master 4.4* was employed for FE analysis of forming limit diagrams. The setting of the numerical simulation is based on the hemispherical punch and different shapes of specimens, as shown in Fig.8. Descriptions of the specimen dimensions and the geometrical model used in the simulation are shown in Table 3 and 4, respectively. The tensile properties of sheet metal were then input into the program and forming limit diagram were generated in *Autoform 4.4* software using Keeler method [16]. *Autoform 4.4* software automatically generates yield surface proposed by Banabic (BBC yield surface) and Hill for sheet materials when anisotropy coefficients and elasto-plastic behavior of sheet are imported. In *Autoform* the use of the shell element for the element formulation is mandatory, and therefore default, for the process steps Drawing, Forming, Bending and Hydroforming. Moreover, since for titanium and ultra high strength steels more complex material laws (for example Barlat or Banabic) are used, *Autoform* uses the implicit integration algorithm which contribution to the total calculation time is substantially smaller.

In this approach, CAD data were modeled in CATIA software first and then imported into *Autoform 4.4* environment. In order to cover full range of the FLD, different specimens with different groove dimensions were modelled to simulate the tension-compression to tension-tension side of the FLD (Fig.8).

For the FE simulation, the punch, holder and die were considered as rigid parts. A displacement rate of 1mm/s was assumed for the hemispherical punch while for the clamping a draw bead with lock mode was selected to ensure pure stretching of the sheet into die cavity. Friction coefficient was taken to be 0.15 between the surfaces. The virtual samples were engraved with the gridded pattern of 3mm diameter circles (Fig.8). Major and minor strains were recorded after each time step to evaluate the numerical FLD.

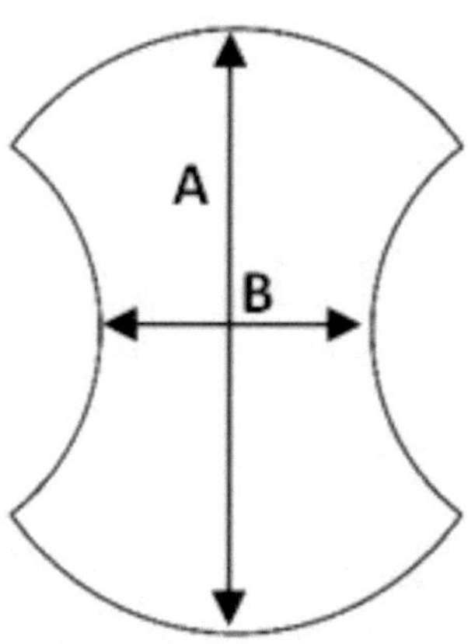

Sample #	A(mm)	B(mm)
1	100	5
2	100	12
3	100	20
4	100	30
5	100	40
6	100	50
7	100	60
8	100	100

Table 3. Dimensions of different FLD samples prepared for FE approach

Process parameter	Value/Type
Punch diameter (mm)	50
Diameter of die opening (mm)	55
Die profile radius (mm)	8
Punch speed (mm/s)	1
Punch travel	Up to rupture
Clamping type	Draw bead (lock mode)

Table 4. Process parameters used for the simulation

Figure 8. Schematic view of the model used in FE analysis as well as gridded sample shapes

5. Results and discussion

5.1. Tensile test

Table 5 illustrates mechanical properties of the tested sheet material deduced from the tensile test at room temperature. Tensile tests results on 1.08 mm thick Ti-6Al-4V sheets show the average values for strain-hardening (n) and anisotropy (r) are (0.151, 3.63). As it can be seen Ti alloy has large plastic strain ratio (r) values. Generally higher strain-hardening exponent (n) delays the onset of instability and this delay, enhances the limiting strain (i.e. a better stretchability and formability is achieved with higher (n) value). Also, increasing plastic strain ratio (r) results in a better resistance to thinning in the thickness direction during drawing which in turn increase the formability of sheet material. On the other hand, high planar anisotropy will bring about earring effect in sheet metal forming processes especially in deep drawing process [19].

Property	Ti-6Al-4V		
	0°	45°	90°
Yield stress, σ_y (MPa)	544	558	571
Ultimate tensile stress, σ_{uts} (MPa)	632	607	629
Work-hardening exponent, n	0.151	0.134	0.167
Hardening coefficient, K (MPa)	975	912.5	1022
Total elongation, δ(%)	30.7	27.2	28.0
Anisotropy factor, r	2.4644	4.1218	3.8292
Normal anisotropy, R_{ave}		3.6343	
Poisson's ratio, υ		0.342	

Table 5. Mechanical properties of tested sheet material obtained in tensile test

5.2. Hydroforming bulge test

Investigation of bursting pressure

As discussed in section 3, in order to discern the bursting pressure of Ti-6Al-4V sheet material, at least three specimens were bulged up to bursting point and average bursting pressure for these alloys was obtained (Table 6). After obtaining burst pressure, test samples were bulged up to 90-95% bursting pressure while the bulge height was being monitored by the indicator. The resulted bulging pressure vs. dome height curves were then extrapolated up to burst pressure by using a third order polynomial approximation. Fig.9 shows bulge pressure versus dome height for the tested material. In this figure, experimentally measured curves along with the extrapolated regions are depicted. In Fig.10 tested samples are shown.

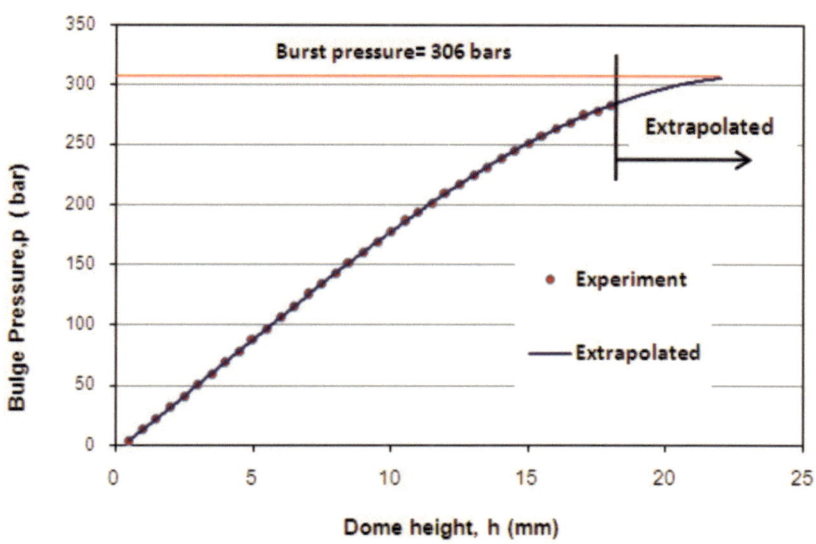

Figure 9. Experimental bulge pressure versus dome height curve for Ti-6Al-4V alloy (the curve is extrapolated)

Burst pressure (Bars)	Ti-6Al-4V
Sample 1	305
Sample 2	307
Sample 3	308
Sample 4	309
Average	306

Table 6. Burst pressure for different samples

Figure 10. Burst and not burst samples of Ti-6Al-4V alloy

Measurements/calculations of bulge radius

In order to measure the bulge radius, several photographs of the bulged samples were taken. The camera was stabilized parallel to the blank. By using Solid Works software, 3-point circle was fitted to the bulge geometry and the radius of the bulge was measured. Fig.11 shows photographs of the bulged samples in stepwise approach. Measured and calculated bulge radii were compared. Among the compared approaches, Panknin's calculations for bulge radius yielded values closer to step-wise bulge results for the alloy tested. Fig.12 shows the comparison of calculated bulge radii with the step-wise measured bulge radius.

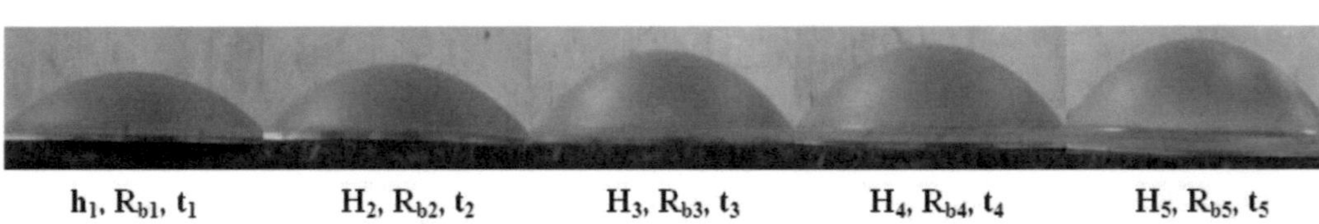

Figure 11. Photography of the bulged sample in stepwise approach

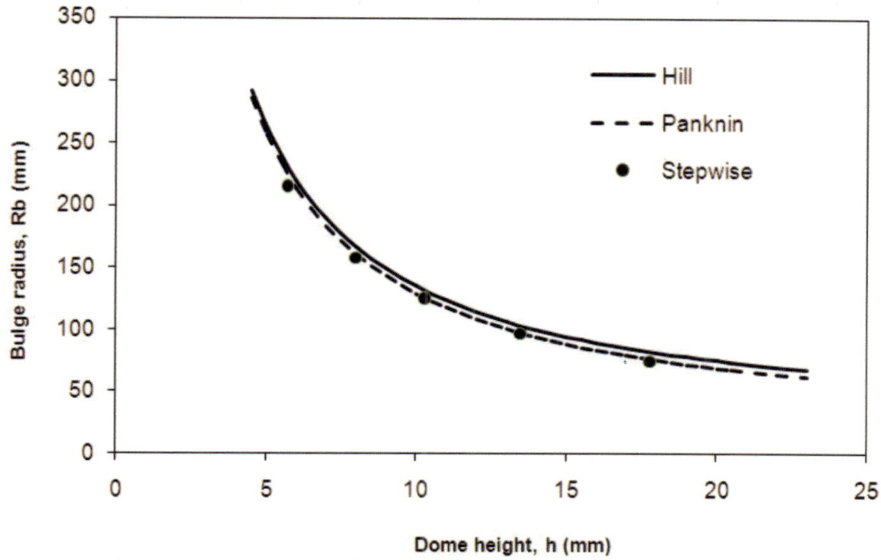

Measurements/calculations of thickness at the dome apex

Figure 12. Comparison of calculated and measured bulge radii for Ti-6Al-4V

Eqs.3-5 were used to calculate the sheet thickness at the dome apex and the results were compared with the step-wise experiments. For measurement of the thickness at the dome apex, a 10mm diameter circle was imprinted on the centre of each bulge sample. After each step, the major and minor diameters of the formed circle were measured by using an accurate caliper (0.05 mm accuracy). Subsequently, the effective strain was obtained for each step at the dome apex by replacing the instantaneous sheet thickness obtained from Eq.26 into Eq.7. The instantaneous thickness of the dome apex was extracted from Eq.8 and is shown as follows:

$$t = \frac{t_0}{e^{(\varepsilon_\theta + \varepsilon_\phi)}} \tag{26}$$

The results show that for calculating the sheet thickness at the dome apex, Kruglov's approach gives best results when compared with step-wise experiments. Fig.13 shows the comparison of calculated sheet thickness at the dome apex with the step-wise measured sheet thickness at the dome of the bulge.

As discussed before (Table 5) higher n-value indicates better stretchability and formability, therefore, for the same bulge height, sheet materials with larger n-values have lower thinning than sheet materials with smaller n-values. Also, when drawing is the deformation mode (e.g. tensile test) r-values strongly influence the thinning process of sheet deformation since higher r-values promote in-plane deformation (ε_1>0, ε_2<0). As a result, Ti-6Al-4V sheet is very resistant to thinning due to the high normal anisotropy during

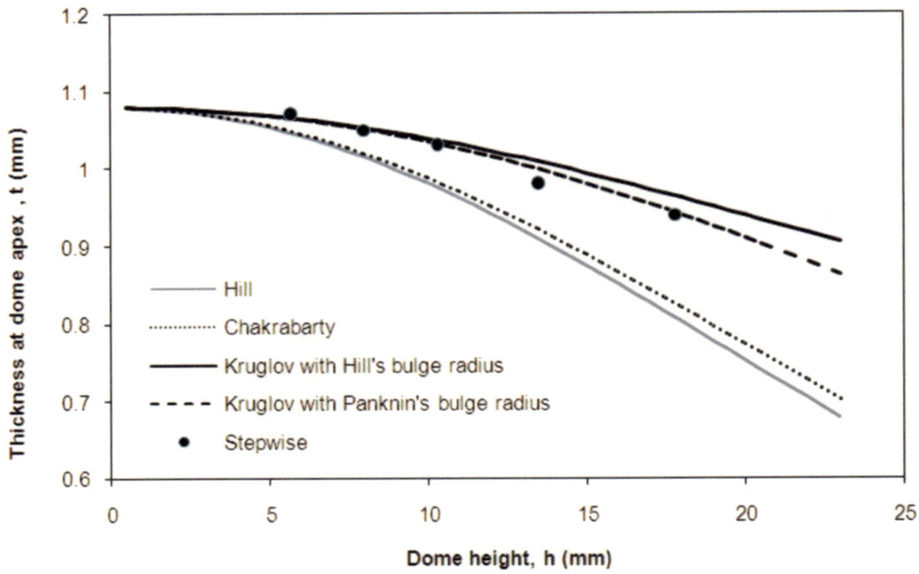

Influence of material properties and anisotropy on material deformation

Figure 13. Comparison of calculated and measured thickness at the dome apex for Ti-6Al-4V alloy

drawing deformations. On the other hand, when stretching is the deformation mode (e.g. bulge test), then both ε_1 and ε_2 are positive and thinning has to occur by constancy of volume (in this case the n-value having a strong influence). This experimental conclusion directly validates the numerical finding obtained by Gutscher et al. [2]. Their FE simulations indicated that anisotropy had very small influence on the correlation between the dome wall thickness at the apex of the dome and the dome height. They also concluded that anisotropy had no significant effect on the radius at the apex of the dome.

Determination of flow stress curves

Several flow stress curves were calculated for Ti-6Al-4V titanium sheet by using several proposed approaches discussed in previous sections. Calculated flow stress curves for titanium alloy were first corrected for anisotropy according to Eqs.9 and 10. Corrected curves are depicted in Fig.14. Step-wise measurements of stress-strain relationships in biaxial test are also shown in the same figure. As it can be seen, step-wise experiments are in good agreement with calculated flow stress curves when Kruglov's and Panknin's approaches are used for dome thickness and bulge radius calculations, respectively.

Fig.15 shows an overall comparison between flow stress curves obtained from tensile (up to instability point) and hydroforming bulge test. A constant scaling parameter was applied to transform biaxial stress-strain curve into effective flow stress curve which can be compared with the uniaxial curve [11]. Kruglov's sheet thickness calculation combined with Panknin's bulge radius calculation was used to obtain these curves. In this figure, tensile curves are depicted along the direction in which the highest elongation was obtained. This comparison

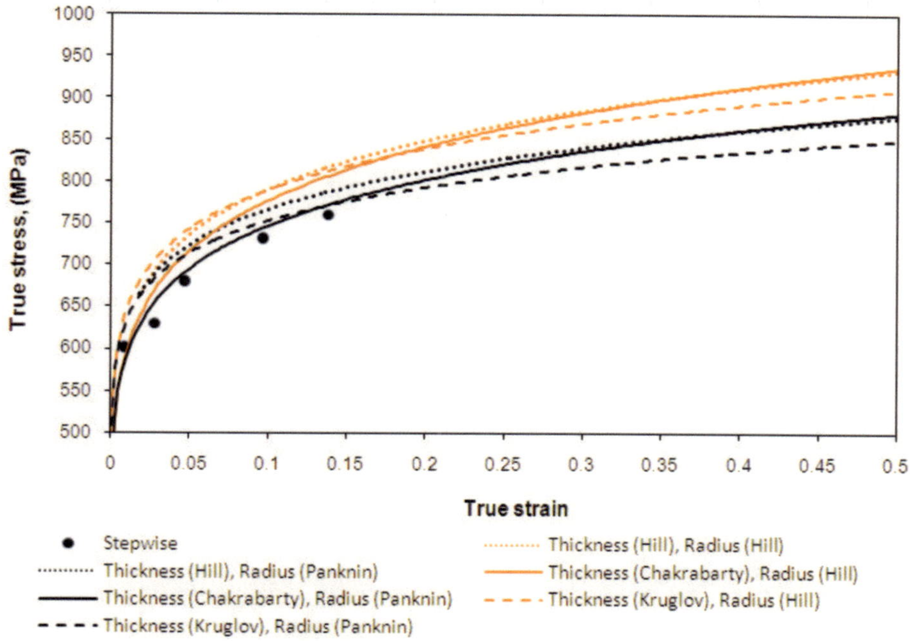

Figure 14. Comparison of measured and calculated flow stress curves for Ti-6Al-4V alloy

indicates, that balanced biaxial bulge test covers larger strain range than tensile test. Under balanced biaxial loading, the theoretical effective strain at instability is twice the instability strain under uniaxial loading. Comparing the data between uniaxial and bulge tests (Fig. 15 and Table 7), it can be seen that strain values obtained in the bulge test are higher than in the tensile test. This is an advantage of the bulge test, especially if the flow stress data is to be used for FE simulation, since no extrapolations is needed as in the case of tensile data. Moreover, in Table 7 it is shown that the percent difference is as high as 504% (for Ti-6Al-4V). This emphasizes the importance of the bulge test because of its capability to provide data for a wider range of strain compared to the traditional tensile test. Also, the constant scaling factor (k_b) [11], which transforms biaxial stress-strain relationships into the effective stress-strain curves, is 0.937 for the tested material.

	Ti-6Al-4V
Maximum true strain that can be obtained in tensile test (up to uniform elongation)	0.115
Maximum true strain obtained in the bulge test	0.58
Maximum difference between tensile test and bulge test, (%)	504

Table 7. Comparison between the maximum true strain in tensile and hydroforming bulge tests

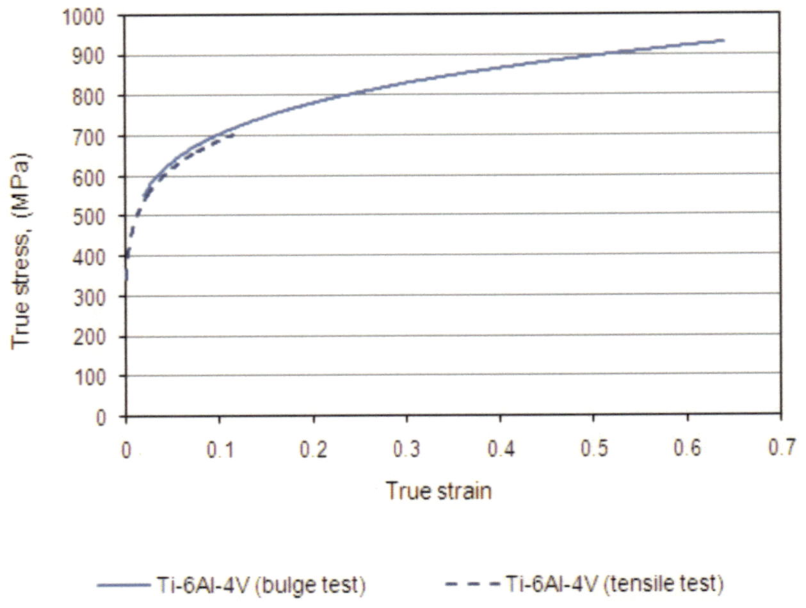

Figure 15. Comparison of effective flow stress curves obtained from tensile and bulge tests

5.3. Determination of the FLD

High resolution photography was employed to measure the diameters of the deformed circles imprinted on the samples (Fig.16). Non-deformed circles were used for calibration of the pictures. As a result, deformed circles were measured using measuring techniques in Solid-works software. Ellipses located in the fractured region, were considered as unsafe points. Likewise, ellipses with one row offset from the fractured region were considered as marginal points and ellipses located in other rows of imprinted grids were considered as safe points. Eqs.27 and 28, were used to obtain true major strain (ε_1) and true minor strain (ε_2) from the measured diameters considering the approach shown in Fig.17.

$$\varepsilon_1^* = Ln\frac{d_1}{d_0} \tag{27}$$

$$\varepsilon_2^* = Ln\frac{d_2}{d_0} \tag{28}$$

Experimental determination and theoretical calculation of the FLD

One of the most important factors for prediction of FLD through the M-K analysis is the applied constitutive yield model. Fig.18 presents the experimental and numerical forming limits for Ti-6Al-4V titanium alloy. For numerical analysis, yield surfaces were described by Hill93 and

BBC2000 yield functions and hardening model was expressed by Swift equation. As it can be observed in Fig.18, although curves predicted using Hill93-Swift model and M-K with BBC2000 are in rather good agreement with the experimental data, the best prediction is obtained when M-K model when the yield surface of Hill93 and initial geometrical defect (f_0) of 0.955, are used. As it can be observed in Fig.18, for Ti-6Al-4V sheet alloy, curve predicted using Hill-Swift model has small deviation from the experimental data from uni-axial region (ρ=-1/2) to equi-biaxial region (ρ=1). Moreover, slight difference in stretching region of the FLD between the experimental data points and the theoretical curve can be seen when BBC2000 is used as the yielding surface for the M-K model.

Note: As discussed by Graf and Hosford [29], applying prestrain in biaxial tension (ρ=1) will decrease the formability if followed by plane strain or biaxial tension. Moreover, for uniaxial tension sample (ρ=-1/2), if both prestrain and final testing with ε_1 are applied normal to the rolling direction, the FLD will be increased for subsequent plane strain and biaxial regions. Furthermore, for plane strain sample (ρ=0) a slight increase of the overall level of the curve is expected when prestrain and final testing with ε_1 are applied normal to the rolling direction.

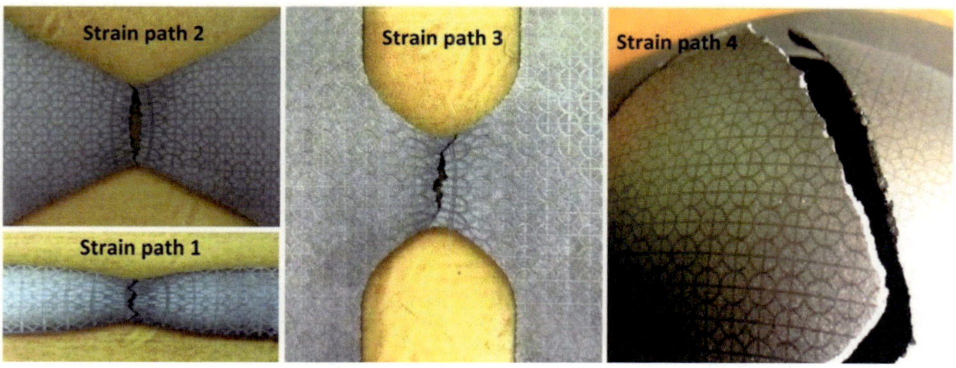

Figure 16. The ruptured tensile and bulge specimens

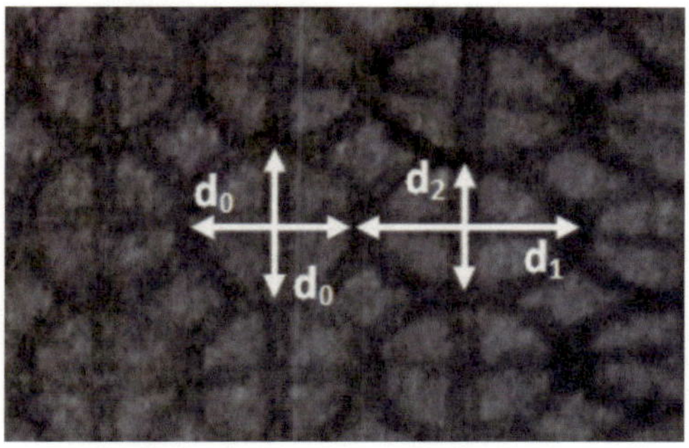

Influence of n-value and r-value on FLD

Figure 17. Deformation of the grid of circles to ellipses.

Generally, there are two material properties which have significant influence on forming limit diagram; the anisotropy and the work-hardening exponent (n-value). R-values less than one ($r_{ave}<1$), will result in the reduction of limit strains in drawing side of the FLD and lower levels for the FLD in plane strain region is expected. For R-values larger than one ($r_{ave}>1$), the opposite trend is expected

Consequently, among the tested sheets, Ti-6Al-4V severely resists to thinning during the sheet metal forming processes.

For most materials, forming limit curve intersects the major strain axis at the point equivalent to n-value. As n-value decreases, the limit strain level decreases. For Ti-6Al-4V, the major strain values are approximately 0.14 and 0.16 at plane strain region of the FLD.

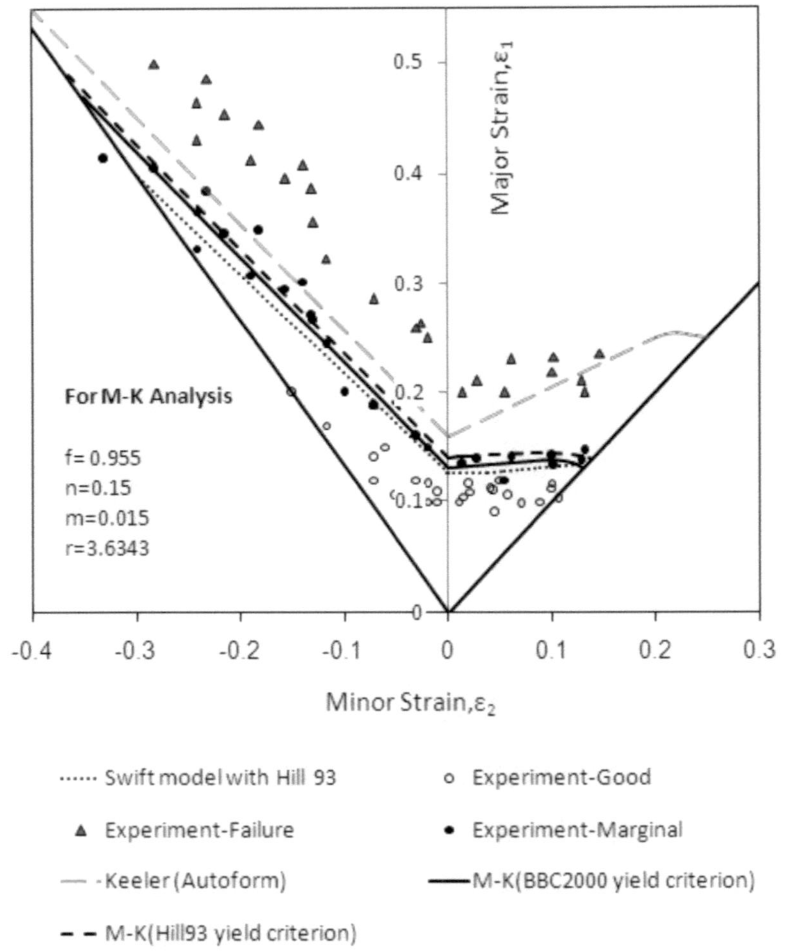

Comparison of experimental FLD with other works

Figure 18. Experimental and calculated forming limit diagram for Ti-6Al-4V alloy.

Fig.19 compares the FLDs for 1.08 mm thick Ti-6Al-4V sheet determined in this study and the same sheet investigated by Djavanroodi and Derogar [19] during hydroforming deep drawing process. Frictional effect between the toolset and the sheet were not considered in their study. Moreover, in hydroforming deep drawing process, the strain rate is far different from the bulge and the tensile test method. Consequently, the experimental procedure and the strain rate are two reasons for this offset between the experimental FLD results for Ti-6Al-4V sheet.

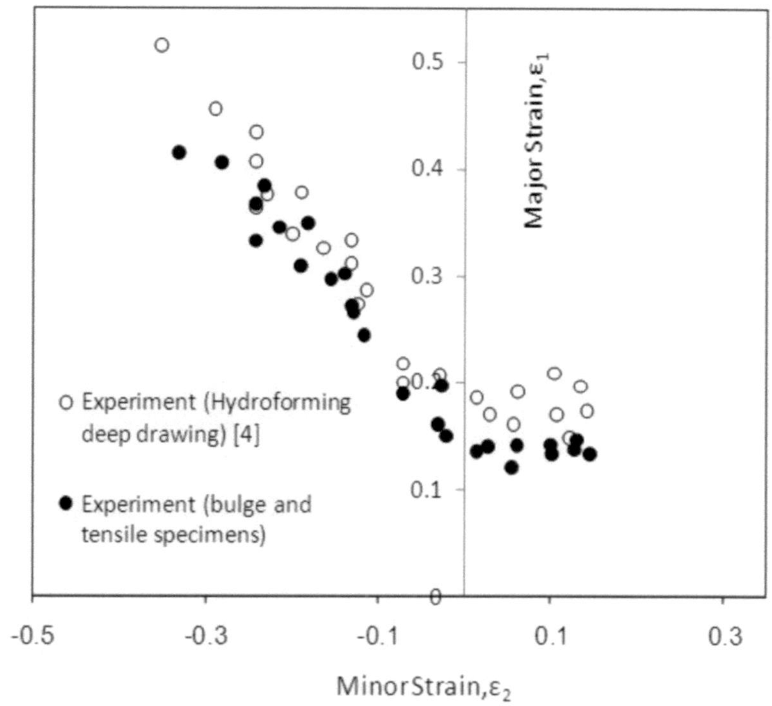

Figure 19. Comparison of forming limit diagrams for Ti-6Al-4V alloy.

5.4. Finite element analysis

Figs.20 and 21 show the numerical results generated by *Autoform 4.4*. The mechanical properties were input into the software, and the yield surface and FLD were generated as shown in Fig.20. Fig.21 shows that engraved circles were deformed and their shapes were changed to ellipses. The major and minor diameters of the ellipses were measured in the software to simulate the FLDs through the FE method. Fig.22 shows the comparison of the experimental and numerical FLDs. As shown in the figure, employing Hill's yield criterion for Ti sheet will result in better prediction of the FLDs compared to experimental data. Predicted FLDs using the industrial sheet metal forming code showed that the shape of the yield loci will have influence on the level of the FLD. Moreover, Fig.22 shows that necking points predicted by Hill's yield criterion and BBC yield criterion stand in good agreement compared to the experimental marginal points, the overall comparison shows a fair agreement between FE results and data obtained from the experiments. The small deviation between numerical and experimental results may be the conclusion of frictional effects between hemispherical punch (in FE simulation) and the sheet metals. While frictional effects remained as an unknown, in order to define the FLDs for different materials, either procedures without frictional effects should be employed or the effect of friction should be taken into account.

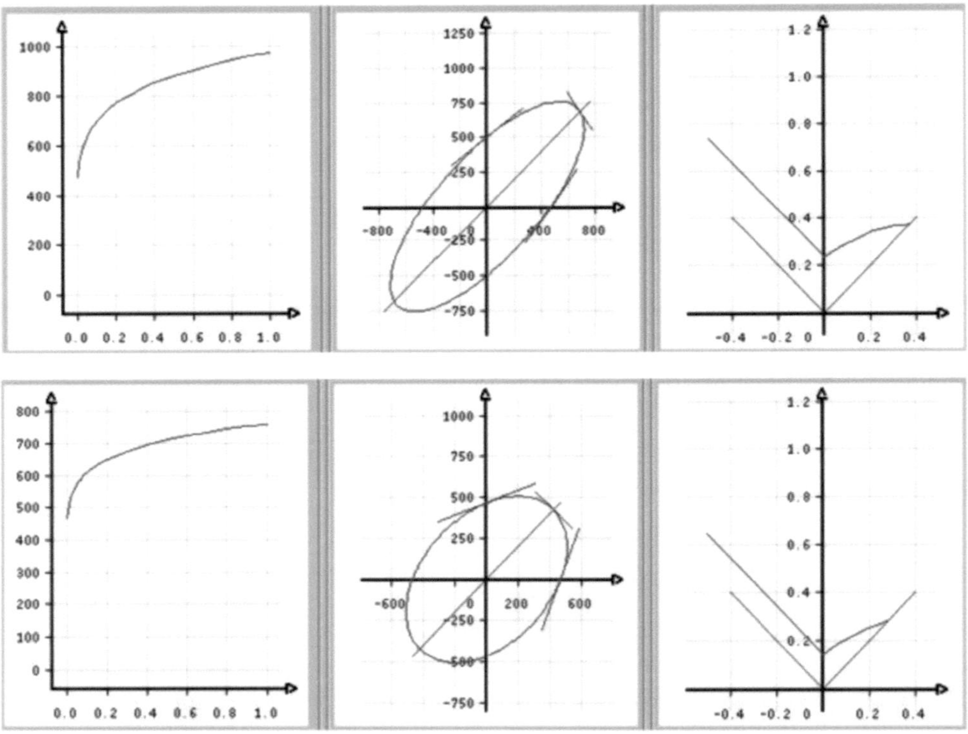

Figure 20. Numerical flow stress, yield locus and FLD generated in *Autoform* software.

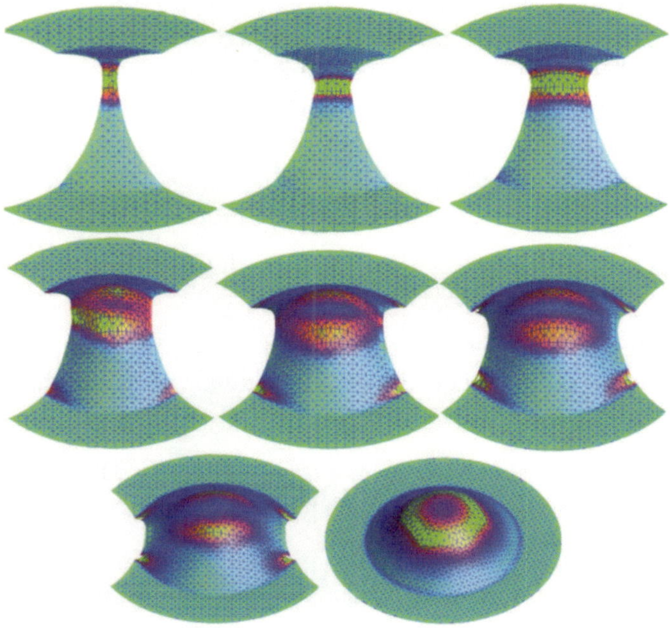

Figure 21. Deformed specimens simulated using *Autoform* software

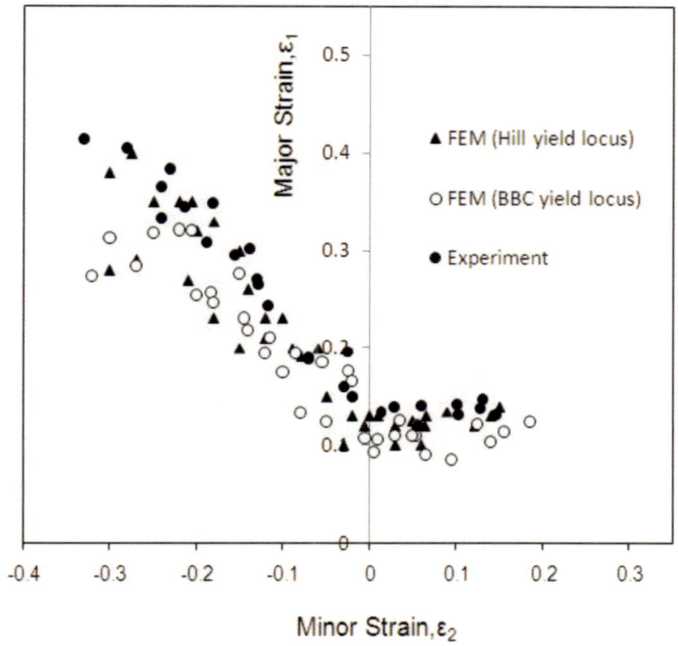

Figure 22. Comparison of the experimental FLD with the ones obtained by FEM for Ti-6Al-4V alloy

Although the forming behavior of materials can be well expressed through uni-axial tensile tests, the theoretical prediction of FLD may still lie in large deviations from the experimentally determined FLDs. This finding proves that suitable theoretical approaches depend not only on the thorough understanding of the forming behavior of materials, but also on the suppositions for yield surfaces as well as material specifications.

6. Conclusions

Based upon the experimental, theoretical and numerical approaches used in this research the following conclusions could be drawn for forming titanium sheet alloys at room temperature:

1. Several approaches were used to generate true stress strain data based on thebulge test. Equations based on Kruglov and Panknin gave the most accurate results comparing to the step-wise experimental measurements.

2. Tensile tests showed that Ti-6Al-4V sheet alloy has large plastic strain ratio (r) values. Generally higher strain-hardening exponent (n) delays the onset of instability and this delay, enhances the limiting strain (i.e. better stretchability is achieved with higher n-value).

3. Investigation of the influence of material anisotropy showed that increase in normal anisotropy would result in lower thickness thinning when drawing is the deformation mode and higher n-value would bring about higher bulge height, lower bulge radius and lower thickness thinning when stretching is the deformation mode.

4. The comparison between uniaxial and balanced biaxial bulge tests indicates that, in bulge test the flow stress curve can be determined up to larger strains than in the tensile test. This is an advantage of the bulge test, especially for metal forming processes in which the state of stress is almost biaxial; using bulge test is a more suitable method for obtaining the flow stress of the above sheet materials for use as an input to Finite Element (FE) simulation models.

5. It was shown that the percent difference for maximum plastic strain obtained from bulge and tensile test is as high as 504% (for Ti-6Al-4V).

6. For Ti-6Al-4V alloy, the best agreement between experimental and theoretical approaches is achieved when the M-K with Hill'93 yield criteria is used. As a result, the overall comparison shows a well agreement between FE results and data obtained from the experiments.

7. The small deviation between numerical and experimental results may be the conclusion of frictional effects between hemispherical punch (in FE simulation) and the sheet metals.

Nomenclature

Nomenclature	Description
d_d	Die diameter
h_b	Bulge height
n	Work hardening exponent
p	Bulge pressure
t_0, t	Initial thickness, instantaneous thickness
c, p, q	Coefficients of Hill'93 yield criterion
a, b, c, k	Material parameters in BBC2000 criterion
d, e, f, g	Anisotropy coefficients of material in BBC2000 yield criterion
m	Strain rate sensitivity factor
t^a_0, t^b_0	Initial thicknesses at homogeneous and grooved region
f_0	Geometrical defect
R_0, R_{45}, R_{90}	Anisotropy coefficients
K	Strength coefficient
R	Average normal anisotropy
R_b	Bulge radius
R_d	Die radius
R_f	Upper die fillet radius
ΔR	Planar anisotropy
$\bar{\sigma}_{isotropic}$	Isotropic effective stress
$\bar{\sigma}_{anisotropic}$	Anisotropic effective stress
$\bar{\sigma}$	Effective stress
σ_1, σ_2	Principle stresses
σ_0, σ_{90}	Yielding stresses obtained from tensile tests
σ_b	Biaxial yield stress
$\varepsilon_\theta, \varepsilon_\phi, \varepsilon_t$	Principal strains
$\varepsilon_1, \varepsilon_2$	Major and minor limit strain
$\dot{\varepsilon}$	Effective strain
ε_0	Pre-strain
$\dot{\varepsilon}$	Strain rate
$d\varepsilon_{2a}, d\varepsilon_{2b}$	Strains parallel to the notch

$\dot{\varepsilon}_{isotropic}$	Isotropic effective strain
$\dot{\varepsilon}_{anisotropic}$	Anisotropic effective strain
ε_l, ε_w	Longitudinal strain, Width strain
Γ, Ψ	Functions of the second and third invariants
ρ	Strain ratio

Author details

F. Djavanroodi[1] and M. Janbakhsh[2*]

*Address all correspondence to: roodi@qec.edu.sa; miladjanbakhsh@mecheng.iust.ac.ir

1 Mechanical Engineering Department, College of Engineering, Qassim University, Saudi Arabia

2 School of Mechanical Engineering, Iran University of Science and Technology, Saudi Arabia

References

[1] Koç M., Billur E., Cora O. N. An experimental study on the comparative assessment of hydraulic bulge test analysis methods. Journal of Materials & Design 2011; 32: 272-281.

[2] Gutscher G., Chih H., Ngaile G., Altan T. Determination of flow stress for sheet metal forming using the viscous pressure bulge (VPB) test. Journal of Materials Processing Technology 2004;146: 1–7.

[3] Nasser A., Yadav A., Pathak P., Altan T. Determination of the flow stress of five AHSS sheet materials (DP 600, DP 780, DP 780-CR, DP 780-HY and TRIP 780) using the uniaxial tensile and viscous pressure bulge (VBP) tests. Journal of Materials Processing Technology 2010; 210: 429–436.

[4] Mahabunphachai S., Koç M. Investigations on forming of aluminum5052 and6061 sheet alloys at warm temperatures. Journal of Materials & Design 2010;31: 2422–2434.

[5] Smith L. M., Wanintrudar C., Yang W., Jiang S. A new experimental approach for obtaining diffuse-strain flow stress curves. Journal of Materials Processing Technology 2009; 209: 3830-3839.

[6] Hill R. A theory of plastic bulging of a metal diaphragm by lateral pressure. Philosophical Magazine 1950; 41(322): 1133–1142.

[7] Dziallach S., Bleck W., Blumbach M., Hallfeldt T. Sheet metal testing and flow curve determination under multiaxial conditions. Advanced Engineering Materials 2007; 9(11): 987–994.

[8] Iguchi T., Yanagimoto J. Measurement of ductile forming limit in non-linear strain paths and anisotropic yield conditions for 11% Cr steel sheets. Iron Steel Inst Jpn International 2007;47(1): 122–130.

[9] Koh CW. Design of a hydraulic bulge test apparatus. MS Thesis. naval architecture and marine engineering; Massachusetts Institute of Technology: 2008

[10] Montay G., François M., Tourneix M., Guelorget B., Vial-Edwards C., Lira I. Strain and strain rate measurement during the bulge test by electronic speckle pattern interferometry. Journal of Materials Processing Technology 2007; 184: 428–35.

[11] Sigvant M., Mattiasson K., Vegter H., Thilderkvist P. A viscous pressure bulge test for the determination of a plastic hardening curve and equibiaxial material data. International Journal of Material Forming 2009; 2: 235–242.

[12] Banabic D., Vulcan M., Siegert K. Bulge testing under constant and variable strain rates of superplastic aluminium alloys. CIRP Annals Manufacturing Technology 2005; 1: 205–208.

[13] Altan T., Palaniswamy H., Bortot P., Mirtsch M., Heidl W., Bechtold A. Determination of sheet material properties using biaxial tests. In Proceedings of the 2nd international conference on accuracy in forming technology; Chemnitz, Germany; 2006

[14] Chamekh A., BelHadjSalah H., Hambli R., Gahbiche A. Inverse identification using the bulge test and artificial neural networks. Journal of Materials Processing Technology 2006; 177: 307–310.

[15] Rees D.W. Plastic flow in the elliptical bulge test. International Journal of Mechanical Sciences 1995; 37(4): 373–389.

[16] Keeler SP. Determination of forming limits in automotive stampings. SAE Technical Paper 1965;42: 683–691.

[17] Goodwin GM. Application of strain analysis to sheet metal forming problems in the press shop. SAE Technical Paper 1968; 60: 764–774.

[18] Janbakhsh M., Djavanroodi F., Riahi M. A comparative study on determination of forming limit diagrams for industrial aluminium sheet alloys considering combined effect of strain path, anisotropy and yield locus. Journal of Strain Analysis for Engineering Design 2012: 47(6): 350-361.

[19] Djavanroodi F, Derogar A. Experimental and numerical evaluation of forming limit diagram for Ti6Al4V titanium and Al6061-T6 aluminum alloys sheets. Journal of Materials & Design 2010;3: 4866-4875.

[20] Rezaee-Bazzaz A., Noori H., Mahmudi R. Calculation of forming limit diagrams using Hill's 1993 yield criterion. International Journal of Mechanical Sciences 2011; 53(4): 262-270.

[21] Verleysen P., Peirs J., Van Slycken J., Faes K., Duchene L. Effect of strain rate on the forming behaviour of sheet metals. Journal of Materials Processing Technology 2011; 211(8): 1457-1564.

[22] Khan A.S., Baig M. Anisotropic responses, constitutive modeling and the effects of strain rate and temperature on the formability of an aluminum alloy. International Journal of Plasticity 2011; 27(4): 522-538.

[23] Palumbo G., Sorgente D., Tricarico L. A numerical and experimental investigation of AZ31 formability at elevated temperatures using a constant strain rate test. Journal of Materials & Design 2010; 31: 1308-1316.

[24] Huang G.S., Zhang H., Gao X.Y, Song B., Zhang L. Forming limit of textured AZ31B magnesium alloy sheet at different temperatures. Transactions of Nonferrous Metals Society of China 2011; 21(4): 836-843.

[25] Inal K.,Neale K.W., Aboutajeddine A. Forming limit comparisons for FCC and BCC sheets. International Journal of Plasticity; 2005; 21(6): 1255-1266.

[26] Shu J., Bi H., Li X., Xu Z., Effect of Ti addition on forming limit diagrams of Nb-bearing ferritic stainless steel. Journal of Materials Processing Technology; 2012; 212(1): 59-65.

[27] Raghavan K.S., Garrison Jr W.M. An investigation of the relative effects of thickness and strength on the formability of steel sheet. Journal of Materials Science Engineering: A; 2010; 527(21-22): 5565-5574.

[28] Uppaluri R., Reddy N.V., P.M. Dixit P.M. An analytical approach for the prediction of forming limit curves subjected to combined strain paths. International Journal of Mechanical Sciences; 2011; 53(5): 365-373.

[29] Graf A, Hosford W. The influence of strain-path changes on forming limit diagrams of Al 6111 T4, International Journal of Mechanical Sciences 1994;36: 897–910.

[30] Tajally M, Emadoddin E. Mechanical and anisotropic behaviors of 7075 aluminum alloy sheets. Journal of Materials & Design 2011; 32(3): 1594-1599.

[31] Hecker SS. Simple technique for determining forming limit curves. SAE Technical Paper 1975;5: 671–676.

[32] Tadros AK, Mellor PB. An experimental study of the in-plane stretching of sheet metal. International Journal of Mechanical Sciences 1978;20: 121–134.

[33] Gronostajski J, Dolny A. Determination of forming limit curves by means of Marciniak punch. IDDRG congress 1980;4: 570–578.

[34] Raghavan KS. A simple technique to generate in-plane forming limit curves and selected applications. Metal Transaction A 1995;26: 2075–2084.

[35] Olsen, T.Y. Machines for ductility testing. Proc American Society of Materials 1920; 20: 398-403.

[36] Keeler, S.P. Plastic instability and fracture in sheet stretched over rigid punches. Phd Thesis, Massachusetts Institute of Technology, Boston, MA: 1961.

[37] Marciniak, Z., Limits of sheet metal formability. (in Polish) Warsaw, WNT, 1971.

[38] Nakazima, K., Kikuma, T., Hasuka, K., Study on the formability of steel sheets. Metallurgical Transaction 1975; 284: 678-680.

[39] Hasek, V. On the strain and stress states in drawing of large unregular sheet metal components (in German). Berichte aus dem Institute fiir Umformtechnik, Universitdit Stuttfart, 1973.

[40] Hoferlin E., Bael A.V., Houtte P.V., Steyaert G., nd Maré C.D. The design of a biaxial tensile test and its use for the validation of crystallographic yield loci, Modelling Simulation, Materials Science Engineering 2000;8: 423.

[41] Koh C.W. Design of a hydraulic bulge test apparatus, M.Sc thesis. Massachusetts Institute of Technology; 2008

[42] Hill R. A theory of yielding and plastic flow of anisotropic metals. Proceeding of Royal Society of London 1948;193A: 197–281.

[43] Swift HW. Plastic instability under plane stress. Journal of Mechanical Physics and Solids 1952;1: 1–18.

[44] Xu S, Weinmann K.L. Prediction of forming limit curves of sheet metal using Hill's 1993 user-friendly yield criterion of anisotropic materials, International Journal of Mechanical Sciences 1998;40: 913–925.

[45] Marciniak Z, Kuczynski K. Limit strains in the processes of stretched-forming sheet metal. International Journal of Mechanical Sciences 1967;9: 609–620.

[46] Marciniak Z, Kuczynski K, Pokora T. Influence of the plastic properties of a material on the forming limit diagram for sheet metal in tension. International Journal of Mechanical Sciences 1973;15: 789–805.

[47] Parmar A, Mellor P.B. Predictions of limit strains in sheet metal using a more general yield criterion, Int International Journal of Mechanical Sciences 1978;20: 385–391.

[48] Lei L, Kim J. Bursting failure prediction in tube hydroforming processes by using rigid-plastic FEM combined with ductile fracture criterion. International Journal of Mechanical Sciences 2002;44: 1411-1428.

[49] Takuda H, Mori K. Prediction of forming limit in deep drawing of Fe/Al laminated composite sheets using ductile fracture. Journal of Materials Processing Technology 1996;60: 291-296.

[50] Takuda H, Mori K. Prediction Prediction of forming limit in bore-expanding of sheet metals using ductile fracture criterion. Journal of Materials Processing Technology 1999;92-93: 433-438.

[51] Takuda H, Mori K. Prediction Finite element analysis of limit strains in biaxial stretching of sheet metals allowing for ductile fracture. International Journal of Machine Tools & Manufacture 2000;42: 785–798.

[52] Fahrettin O, Daeyong Li. Analysis of forming limits using ductile fracture criteria. Journal of Materials Processing Technology 2004;147: 397-404.

[53] Kumar S, Date P.P, Narasimhan K. A new criterion to predict necking failure under biaxial stretching. Journal of Materials Processing Technology 1994;45: 583-588.

[54] J.D. Bressan, J.A. Williams. The use of a shear instability criterion to predict local necking in sheet metal deformation, International Journal of Mechanical Sciences 1983;25: 155–168.

[55] Hill R. A user-friendly theory of orthotropic plasticity in sheet metals. International Journal of Mechanical Sciences 1993;35(1): 19–25.

[56] Banabic D., Kuwabara T., Balan T., Comsa D.S. An anisotropic yield criterion for sheet metals Journal of Materials Processing Technology 2004; 157-158: 462-465.

[57] Panknin W. The Hydraulic Bulge Test and the Determination of the Flow Stress Curves. Phd thesis, Institute for Metal Forming Technology, University of Stuttgart, Germany, 1959

[58] Chakrabarty J., Alexander J.M. Hydrostatic bulging of circular diaphragms. Journal of Strain Analysis 1970; 5: 155–161.

[59] Kruglov A.A., Enikeev F.U., Lutfullin RYa. Superplastic forming of a spherical shell out a welded envelope. Materials Science Engineering A 2002; 323: 416–426.

[60] Koç M., Aueulan Y., Altan T. On the characteristics of tubular materials for hydroforming – experimentation and analysis. International Journal of Machine Tools and Manufacture 2001; 41(5): 761–772.

[61] Butuc M.C. Forming limit diagrams. Definition of plastic instability criteria, Ph.D thesis, University do Porto; 2004.

[62] Banabic D. Limit strains in the sheet metals by using the new Hill's yield criterion (1993). Journal of Materials Processing Technology 1999;92-93: 429–432.

[63] Wang L., Lee T.C. The effect of yield criteria on the forming limit curve prediction and the deep drawing process simulation, International Journal of Machine Tools & Manuf.; 2006; 46: 988-995.

[64] ASTM Committee E28/Subcommittee E28.02, Standard Test Method for Plastic Strain Ratio r for Sheet Metal, ASTM E517-00; 2006

[65] Bhagat, A. N. , Singh, Avtar , Gope, N. and Venugopalan T. Development of Cold-Rolled High- Strength Formable Steel for Automotive Applications. Materials and Manufacturing Processes, 2010;25(1): 202-205.

[66] Panda Sushanta Kumar, Kumar Ravi D. Experimental and numerical studies on the forming behavior of tailor welded steel sheets in biaxial stretch forming. Journal of Materials & Design 2010;31: 1365-1383.

[67] Ko D.C, Cha S.H, Lee S.K, Lee C.J, Kim B.M. Application of a feasible formability diagram for the effective design in stamping processes of automotive panels. Journal of Materials & Design 2010;31: 1262-1275.

Chapter 6

Physicochemical and Radiation Modification of Titanium Alloys Structure

Kanat M. Mukashev and Farid F. Umarov

Additional information is available at the end of the chapter

http://dx.doi.org/10.5772/55485

1. Introduction

Intense development of science and technology with ever-increasing needs in new materials with the unique properties requires implementation of careful research in this area. Development and production of new types of materials is always related to the enormous costs and solving new technical problems of analytical and experimental nature. Recently a large class of the model alloys on metallic base, which meets various demands, has been created. A special place among them is occupied by the metals and alloys that undergo phase transformations. The requirements for the radiation resistance of these materials are of extraordinary importance. Investigation of the fundamental properties of materials that determine their physical, chemical, mechanical, technological, operational and other characteristics enables one to establish a field of their rational application with maximal efficiency.

The attention of the researchers should be drawn to investigation of the structure transformations in crystals, especially, their electronic and defect structure as well as their role in the process of formation of the material's final physical properties. Almost all of the material's properties are related to its electronic structure, and the constancy of the structure under external exposure determines stability of the main characteristics and can serve as a principal indicator of materials radiation resistance.

The problems of nuclear and thermonuclear power pose an urgent demand on continuous and wide range investigation of interaction processes of nuclear radiation with metallic materials, along with subsequent modification of their structure. For authentic establishing of common regularities of the observed phenomena, deep understanding of the processes of nucleation, formation and subsequent evolution and modification of the metals and alloys defect structure is crucial.

In spite of considerable amount of realized investigations, the analysis of the obtained results justifies the following findings: by the time of preparation of this work, the lack of the systematic information was experienced about the character of the radiation damageability of some perspective constructional refractory metals and their alloys, which firstly undergo polymorphous or phase transformations; influence of the type and concentration of alloying elements on the character of the structural disturbances at plastic deformation and radiation exposure in conditions of vacancy and vacancy-impurity complexes formation, packing defects, dislocation loops subject to material history, fluence, energy, flux, nature of ionizing radiation, temperature of irradiation and postradiational annealing.

There was no a thorough research of the influence of preliminary thermochemical treatment, including hydrogen and other atomic gases saturation and cyclic thermal shocks with an account of reconstruction of electron structure and density of pulse distribution of electrons in the field of defect production on the materials' final properties. Availability of such data would complete a full picture of purposeful properties changes and make possible working the materials with predetermined properties. This problem definition caused by demands of the state-of-the-art science and technology appears to be strategically important area of research in the fields of physics of metals, physics of radiation damage and radiative study of materials.

Therefore, the main goal of the present work, which is based on the authors' own research, is investigation and establishment of regularities of the electron structure alteration and its correlation with different titanium alloys crystal lattice defects created as a result of deformation radiation and complex thermochemical treatment.

2. Experimental and software-supported investigations

Since science of metals is generally experimental science, the depth, objectivity and reliability of our understanding of investigated phenomena related to materials electronic and defect structure are determined by capacity of the technical means and methods of investigation used in order to solve the problem.

The method of positron spectroscopy is the most important instrument of investigation in this case. Not only did the relativistic quantum mechanical theory developed by Dirac (1928) explain the main properties of electron and obtain the right values of its spin and magnetic moment, but also it determined the positron existence probability. Positron is the antiparticle of electron with the mass $m_{e+} = m_{e-} = 9.1\ 10^{-28}$ g same as electron's, rest energy $m_0c^2 = 0.511$ MeV and elemental, but opposite in sign, electron charge $e = 1.6\ 10^{-19}$ K and spin $S = 1/2h$.

Natural positron sources normally do not exist. Therefore positrons are usually obtained from nuclear reactions in different nuclear power plants. The principal criteria for choosing positron sources are the cost and half-life period. The most widespread is the sodium isotope ^{22}Na obtained from ^{25}Mg by reaction (p,α). It is convenient in all respects and easy-to-use in the positron spectroscopy experiments, as well as for angular distributions measurements, Doppler broadening of annihilation line, positrons life time and counting rate of 3γ – coinci-

http://dx.doi.org/10.5772/55485

dence. ^{22}Na nuclear decay occurs by the following scheme: ${}^{22}_{11}Na \rightarrow {}^{22}_{10}Ne + e^{+} + \gamma_{Я}$. In this nuclear decay reaction the ^{22}Na nuclear is produced in an excited state with the time of life less than 10^{-12} s. At the return to the ground state it emits the nuclear quantum with energy E=1.28 MeV, which effectively testifies positron production.

The essence of using positrons for solid structure probing is explained in the following. A positron emitted by a source while penetrating in solid to a certain depth subject to energy, experiences numerous collisions with the atoms of the solid, and consequently this positron gradually loses its velocity and at the end gains energy that corresponds to environment's absolute temperature: $E_0 = kT$=0.025 eV, where k is the Boltzmann constant. This process is referred to as the positron thermolysis. The fundamental result of this phenomenon is the positron thermalization time, during which the positron dissipates its initial energy. Its calculated value is 3 10^{-12} s. [1].

Positron thermolysis process occurs during the time, which is considerably shorter than its life time before annihilation. This circumstance serves as grounds for using positrons in order to study the properties of condensed matters, because the conduction electrons, with which positron interacts, occupy the energetic band of the range of several electron-volt and more importantly positron does not contribute to the total pulse and energy of the pair and hence can be neglected. Therefore, the information, which is carried by the positron annihilation photons, corresponds to solid electrons state, in which positron's thermolysis and interaction and annihilation processes have occurred.

Annihilation is the act of mutual destruction of a particle and its appropriate antiparticle. While no absolute destruction of either matter or energy occurs, instead there is a mutual transformation of particles and energy transitions from one form to another.

Due to the law of charge parity a positron in singlet state ($1S_0$) decays with emission of even number (usually two) gamma-ray quanta. A positron in triplet state annihilates with emission of odd number (usually three) photons. The probability of the 3γ - process is lower by more than two orders than the probability of the 2γ - process. Therefore, all basic research that is oriented towards studying properties of condensed state properties is performed around this phenomenon.

If an annihilation pair is found in the state of rest in center-of-mass system (v=0), then in laboratory system of coordinates two photons would be emitted strictly in opposite directions at sinθ=0 (Fig.1a). As a result of interaction with the medium's electrons and phonons, positron completely thermalizes and in essence is in state of rest. However, we cannot state the same about electron, the other immediate participant of the annihilation process. At the same time pulse transverse component leads to deflection of $\gamma_{1\text{ and }}\gamma_2$ photons from collinearity:

$$P_Z = m_0 c\theta_z; \tag{1}$$

This circumstance is initiating development of the method of measuring angular distribution of annihilated photons (ADAP). The purpose of the method is to obtain information about electrons distribution function in momentum space. To this effect, it is supposed that length of the detector (ℓ), which registers the annihilated photons, must be much longer than its width (δ), and substantially smaller than distance (L) from positron source to detector. These facts correspond to so-called long-slot geometry of experiment, which is schematically represented in Fig. 1a.

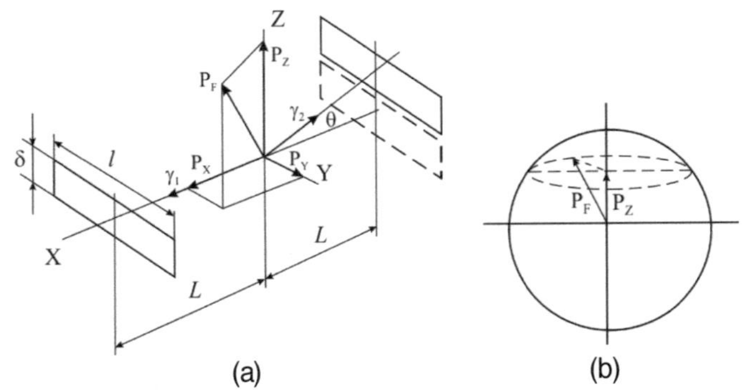

Figure 1. Line and slot geometry circuit with pulse decomposition e^{-}-e^{+} - pair on components (a) and Fermi surface cross-section for gas of free electrons (b)

At rest positron the impulse of annihilated photons is defined by electron impulse. The latter is uniformly distributed on whole Fermi sphere for ideal gas of electrons [2,3]. Therefore, ADAP measuring boils down to choosing thin sphere layer on distance P_z from its center located perpendicularly of this component of impulse (Fig. 1b). In this case angular correlation spectrum $N(\theta)$ must be of shape of reverse parabola, which mathematically can be described by the following equation:

$$N_P(\theta) = N(0)(\theta_F^2 - \theta^2) \text{ for } \theta \le \theta_F \tag{2}$$

This distribution vanishes outside θ_F and this area corresponds to the boundary Fermi momentum$P_F = m_o c\ \theta_F$. Besides the parabolic component the spectrum also contains a wide angular component caused by positrons annihilation with inner electrons of ion core with the impulse that considerably exceeds the Fermi momentum. The regularities of positron annihilation in this case have sufficiently reasonable description by Gaussian function:

$$N_g(0) = N_g(0)\exp\left(-\theta^2/\theta_g\right), \tag{3}$$

where θ_g is a Gaussian parameter and determines the penetration depth of positron's wave functions into the ion core. Hence, the ADAP general curve for any materials can be presented as the following:

http://dx.doi.org/10.5772/55485

$$N(\theta) = N_P(0)(\theta_F^2 - \theta^2) f(\theta) + N_g(0)\exp(-\theta^2 / \theta_g) + N_0 \tag{4}$$

The normalizing factor $f(\theta)$ in this equation takes only the following values:

$$f(\theta) = \begin{matrix} 1 & \text{ï ðè} & |\theta| \le |\theta_F| \\ 0 & \text{ï ðè} & |\theta| > |\theta_F| \end{matrix} \tag{5}$$

The constant factors $N_p(0)$, $N_g(0)$ and N_0 in equation (4) define the intensity of the Gaussian parabola at $\theta=0$ and the background level of random coincidence, respectively. Besides the angular distribution there also exist other characteristics, which describe the regularities of electron-positron annihilation phenomena (EPA). The most important among them are the positron lifetime and Doppler broadening of annihilation line (DBAL). The logical interrelation of these three methods of positron annihilation is schematically depicted in Fig.2

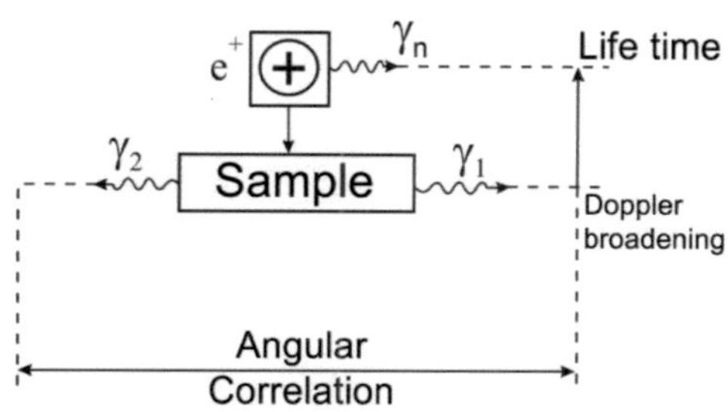

Figure 2. Schematic diagram of different methods of electron-positron annihilation (EPA)

The positrons annihilation process in solids can be described by a set of parameters. But, the most informative for material properties characteristics are those, which successfully fit into different physical regularities, i.e. those that carry in themselves one or another physical meaning. One of these parameters can be the values of probability of positrons annihilation with free and bound electrons. These parameters are derived from processing of experimental angular distributions spectra of annihilation emission (4). The area under each component (S_P, S_g) is usually defined by integration. Knowing the total area under the whole curve $S_0 = \int_{-\infty}^{+\infty} N(\theta) d\theta$, we can calculate the positron annihilation probability with free electrons and electrons of ion core, respectively:

$$WP = SP / S0, WG = Sg / S0, \tag{6}$$

as well as the redistribution of positron annihilation probability between the free electrons and the electrons of ion core:

$$F = WP / WG = SP / Sg. \tag{7}$$

The example of decomposition of experimental spectra to components is shown in Fig. 3. Due to parabolic component spreading in the $\theta = \theta_F$ region, the Fermi angle value is usually determined by extrapolation.

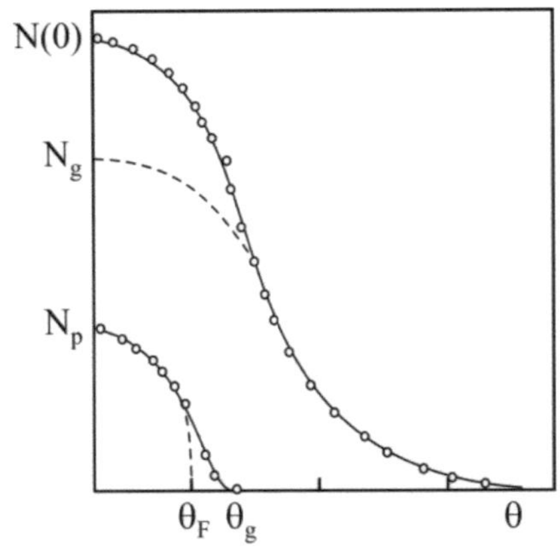

Figure 3. The decomposition of the angular correlation spectra into components

The changes in the investigated material structure are by all means reflected on the spectra form and lead to redistribution of positron annihilation probabilities. In this case after normalization to a single area, they can be built on one axis for comparison purposes (Fig. 4).

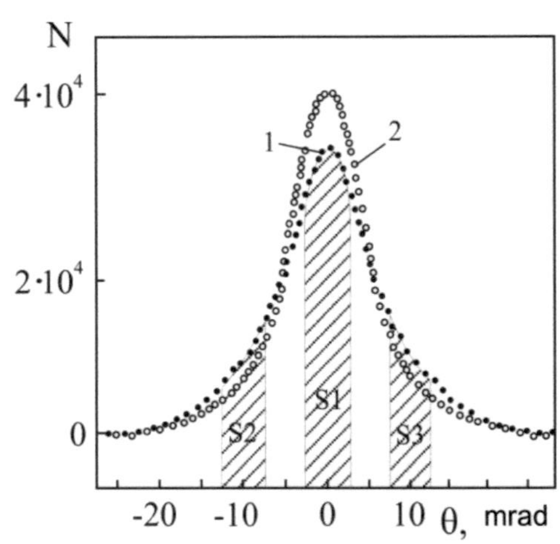

Figure 4. The APAD spectra normalized to a single area for annealed (1) and deformed (2) titanium

While comparing results of one set of measurements, which are related to thermal, deformative or radiative influences on the investigated materials, one can use a special configuration parameter sensible to the presence of only one kind of defect in the crystal [4]:

$$R_{\tilde{N}} = \left| \frac{N_V^t - N_V^f}{N_c^t - N_c^f} \right|, \tag{8}$$

where N is a counting rate value in the structurally sensitive region of the annihilated photons angular distribution spectrum; subscripts v and c of N refer to a positron annihilation with free and core electrons, respectively; superscripts t and f correspond to positron annihilation from the trapped and free states. It is assumed that this parameter does not depend on defects concentration in the material and is determined by its structure only.

The basic specifications of the experimental spectrometer of annihilated photons angular distributions with line and slot geometry are the following:

- Angular resolution of the setup changes within the range of 0.5–1.5 mrad.
- The time resolution on the fast channel equals to 100 ns and on the slow channel ranges within the interval of 0.3–1.0μs.
- The movable detector step width is set stepwise by 0.25, 0.5 and 1.0 mrad; in all the setup permits to measure up to 50 values of coinciding gamma photons intensity in one direction from the spectrum maximum position.
- The counting rate instability in the course of three days of continuous work does not exceed two standard deviations.
- The allowed maximum intensity of incoming information on the slow channel is no worse than 3×10^5 s^{-1}.
- The maximal vacuum in the measuring chamber is no worse than 10^{-4} Pa at the temperature of 300 – 1100 K.
- The positrons source activity of $_{22}Na$ is 3.7×10^8 Bk (10 mKi).

The reliability of positron investigations results depends on a number of reasons which are as far as possible taken into account during the process of experimental investigations.

The preparation of investigated objects of different composition was realized in the high temperature electroarc furnace on a copper bottom with nonexpendable electrodes. After batch charging and before alloying, a vacuum of ~10^{-2} Pa was created in the furnace. After that a high purity argon was introduced into the furnace, in which all the processes of melting were conducted in this atmosphere. For homogeneity, ingots were repeatedly melted (up to 5–6 times).The finished ingots were rolled at a temperature of 900^{0}C up to 1–2mm strips and then annealed at 10^{-5} Pa vacuum at 900 ^{0}C during 2 hours. The annealed samples were prepared from 1mm strips, which repeatedly underwent plastic deformation ($\varepsilon = 50\%$) by rolling at room temperature. The true content of the components in the check samples of the material was determined by chemical and spectral analysis methods. Then the surfaces of these samples were thoroughly burnished, and the samples were polished electrochemically in the solution of the following composition: HF – one part, HNO_3 – three parts and H_2O – two parts. The prepared samples were flushed with flowing water and wiped with alcohol. The temperature measurement of samples at irradiation and annealing was performed by thermometry methods. All measurement processes of annihilation characteristics of investigated materials after different kinds of influences (plastic deformation, radiative and other complex physico-chemical and thermal treatments) were conducted at room temperature.

2. Modification of titanium alloys defect structure by plastic deformation method

The progress of modern engineering and technology is closely related to the achievements in science of metals, which before taking a specified form and properties usually undergo plastic deformation. Not only is the deformation process one of the effective methods for giving the required form to a material but it is also an important means for modifying its structure and properties. Yet defect formation and defects influence on metals physico-mechanical properties is one of the important problems in metal physics. For investigation of modification processes in metals structures the titanium binary alloys, alloyed with Zr, Al, Sn, V, Ge and In within the range of solid solution, were prepared. The elements content in alloys was defined more precisely by chemical and spectral analysis.

Zirconium Zr is an analog of Ti and forms with it a substitutional solid solution of complete solubility. As far as concentration of Zr is increased the temperature of the allotropic transformation of Ti slightly drops and reaches the minimum at equiatomic ratio (50 at.% and 545ºC). Thus, Zr is a weak β–stabilizer for Ti. Usually β–phase is not preserved in this system at the room temperature. Al is a substitutional element for Ti with limited solubility in α- and β-phase at presence of peritectoid breakup of the β–solid solution. The Ti-Al system plays the same role as the Fe-C system for steels in physical metallurgy.

While alloying Ti by Sn, the eutectic systems are formed. Sn forms the system with limited solubility of alloying elements at presence of the eutectic breakup of β–solid solution. Sn considerably differs from Ti by its properties and it is restrictedly soluble in both Ti modifications. Ti-In is one of those systems that are most insufficiently explored due to considerable difficulties related to preparation of alloys. In slightly reduces the temperature of alloy polymorphic transformation and is therefore a weak β–stabilizer.

When alloying titanium with vanadium a solid solution in β-Ti is formed, with complete solubility. Ti-V phase diagram strongly depends on the method of obtaining Ti (iodide, hydride-calcic or magnesium-thermic). The V solubility in α–Ti at 650ºC does not exceed 3.5 weight %. Alloying with V leads to Ti lattice spacing decrease, therefore the c/αratio consecutively decreases with increase of V content in alloy.

2.1. Structural transformations in plastically deformed alloys of the Ti-Zr system

Influence of plastic deformation on the structural damages formation was investigated on the alloys that contained 0; 2.7; 8.3; 17.0; 22.0 and 39.0 at.% Zr. Plastic deformation ε = 50% considerably changes shape of the curves by decreasing width on the half height (FWHM) and increase of counting rate at maximum of $N(0)$ relative to the initial (annealed) state, which is the effect of crystal structure defects occurrence in the material (Fig.4). For interpretation of investigation results the above mentioned annihilation parameters $F = S_P/S_g$, $\theta_F = P_F/mc$ and their relative changes ΔF and $\Delta\theta_F$ were used.

In order to establish regularities of changes of annihilation parameters depending on the structure damages level, Ti and its alloy Ti – 2.7 at.% Zr underwent plastic deformation by different degrees in the range from ε = 5 up to 80 %. It is determined that main changes of the annihilation parameter occur when ε takes values up to 30%, whereupon it reaches saturation. Though for Ti–2.7 at.% Zr alloy it reaches saturation considerably quicker than for pure Ti (Fig. 5). Evidently it can be stated that with an increase of deformation degree the defects concentration in metals also increases and at certain level virtually all positrons are captured and annihilated in defects. In such state the annihilation parameters are typical for deformation defects. In other words, in the conditions of ultimate strains the EPA characteristics contain information about structure of the crystal regions, in which the vast majority of positrons annihilation occurs.

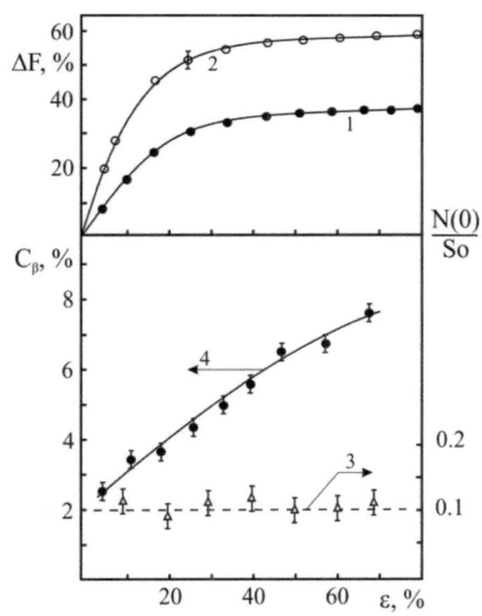

Figure 5. The influence of deformation degree and alloying element concentration on titanium alloys structural-sensitive characteristics: 1-Ti; 2- Ti-2.7 at.% Zr; 3- $N(0)/S_0$ relationship; 4- the β-phase content in alloy

In order to determine the phase composition, titanium and Ti–2.7at.% Zr alloy underwent X-ray analysis using the DRON-2 diffractometer in the filtered CuK_α radiation with the help of the special methods of accuracy enhancement. Then the β–phase concentration in the lattice structure can be determined by the equation [5]:

$$C_{\beta} = \frac{100\%}{1+2.33(I_{\alpha}/I_{\beta})}, \tag{9}$$

where I_α and I_β are X-ray radiation integral intensity for α- and β–phases of the alloy, respectively. It has been established that the cold deformed Ti at any deformation degrees has only a single-phase structure. Also after plastic deformation in the alloy there is a two-phase α+β microstructure is observed, where the β–phase content changes monotonically with the deformation degree (Table 1 and Fig.5).

E (%)	0	5	10	20	30	40	50	60	70	80
C_β (%)	-	2,66	3,61	3,76	4,85	5,90	6,96	7,25	8,05	8,77

Accuracy ±0,05

Table 1. The β–phase content in Ti–2.7at.% Zr alloy

Hence, one can do the following conclusion. For those alloys that undergo transformation, plastic deformation at room temperature initiates phase transformation since energy rise in the crystal introduced by the defects may thereby decrease. In this case, the boundary with matrix regions of new phase nucleus should be assumed as the most likely positrons capture centers.

In such metals as Ti and Zr, in which the phase transformations from HCP to BCC-structure occur at relatively low temperatures, the packing defect formation energy in prismatic plane must be small [6]. With this in mind one cannot help noticing the nature of dependence of the relative change $\Delta F = (F_{def} - F_{rel})/F_{rel}$ parameter on alloys composition (Fig.6).

The maximum changes ΔF are observed for alloys with 2.7 and 3.9 at.% Zr, while for other concentrations this value is significantly lower. By individual cases, possible role of other factors is indicated by this behavior of the parameter F; one of such factors can be interaction of localized-in-defects positrons with lattice instability. In other words, it is pertaining to the different degrees of alloys lattice stability towards transformation initiation. It is evident that the maximum deviation of the parameter F related to Zr concentration nearly corresponds to the minimum value of the α→β transformation temperature.

Experimentally obtained value of $\Delta\theta_F=(\theta_\varepsilon-\theta_{F0})/\theta_{F0}$ parameter reaches ~9,5%, which is higher by order of magnitude than expected. This testifies that the positrons are annihilated in the defect regions, in which electron density is significantly lower than in the matrix. Thus the most likely positrons capture centers in this case are the new phase regions on its boundaries with the matrix. The structure of these new phase regions differs by far from that of the matrix.

http://dx.doi.org/10.5772/55485

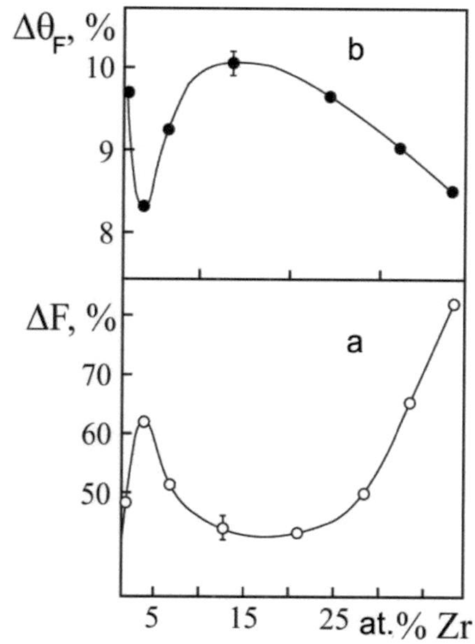

Figure 6. Concentration dependencies of annihilation characteristics for Ti-Zr system alloys.

Therefore, one should consider two main factors that are responsible for initiation of the polymorphic transformation in the Ti-Zr alloys: plastic deformation, which leads to the formation of packing defect with the BCC-phase structure in the HCP-phase matrix, and positron interaction with lattice instability. Separation of contribution pertaining to each factor to annihilation parameters change is as yet an impracticable problem for employed experiment conditions.

2.2. Structure modification in the Ti-Al and Ti-In alloys

In order to obtain additional information about the nature of positrons interaction with the structure damages in the plastically deformed metals the second group of binary titanium alloys was prepared. These alloys contain alloying elements from the III group of the periodic system, namely Al in concentrations 0; 5.2; 10.2; 12.5 and 16.5 at.%, and also In with concentrations 0; 1.4; 2.9; 5.1; 8.5 and 10.3 at.%. The maximum concentration of alloying element in each of these systems meets the requirements of mandatory occurrence of alloys in the solid solution region, where the chemical compounds formation is ruled out in advance.

All alloy samples with the specified concentrations of alloying elements were deformed by the cold rolling method by ε =50%. Concentration dependencies of the ΔF and $\Delta\theta_F$ annihilation parameters for the investigated materials are presented in Fig. 7. One can see that relative changes of the F parameter for these alloys are also considerable as in the previous case and reach 130% for Ti–5.2 Al alloy and 250% for Ti–1.4 In alloy. At the same time the Fermi angle decrease for the first case reaches $\Delta\theta$ = 17% and 12.7% - for the second case. For Ti-In system the ΔF concentration dependence is of complex nature with two maximum points with In concentrations 1.4 and 7.4 at.%.

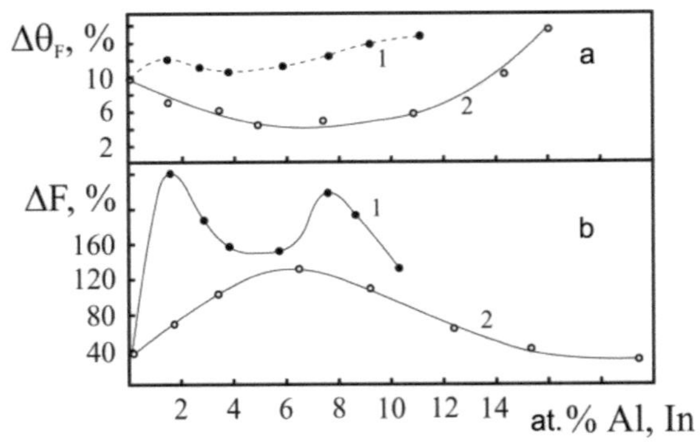

Figure 7. The concentration dependencies of annihilation parameters for Ti-In (1) and Ti-Al (2) deformed alloys systems.

If a strongly deformed alloy is to be considered as a two-phase system, then on the basis of positrons capture models one can define approximate bulk defect size therein: R_V = 10 Å. Therefore, one can assume that strong deformation of titanium alloys is accompanied, along with the new phase, by the formation of vacancy clusters and their aggregations, which can serve as deep potential wells for positrons.

2.3. Peculiarities of positrons annihilation in the deformed Ti-Sn alloys

These alloys contained the following concentrations of alloying element: 0; 1.2; 2.5; 4.3; 6.2 and 7.6 at.% Sn. All alloys were deformed by cold rolling by $\varepsilon = 50$ %. The investigation results are presented in Fig.8. The observed changes of parameters in this case, as in the earlier one, are substantial and possess a complex nature dependent on alloy composition. Thus, the relative change of annihilation probability ΔF for Ti–1.2 at.% Sn alloy passes through maximum and reaches 135 % and then reaches minimum value ($\Delta F = 85$ %) at the concentration of 6.2 at. % Sn. For comparison, in the case of Ti-Ge alloys system this factor also passes through maximum at the concentration of 0.8 at.% Ge, but with ΔF =185 %.

One can make yet another observation which is typical for these alloys: the nature of change of the alloying elements parameters at small concentrations roughly coincide, whereas maximum decrease of the Fermi angle ($\Delta\theta_F = -17\%$) is detected for the Ti–1.5 at.% Ge alloy. In other words, it is facing a certain shift between maximum locations for two dependencies of one system of alloys.

Based on abnormally large changes of F and θ_F annihilation parameters one can make an assumption that the positrons capture centers structure conform to the regions with average electron density, which is considerably lower than for general vacancy-dislocation defects. Since the deformation causes a considerable increase of the parameter F with the respective decease of the Fermi momentum, this indicates that with general decrease of average electron density in these defects the contribution of ion core electrons to the EPA process also decreases. This means that the annihilation occurs mainly with free electrons in the defects.

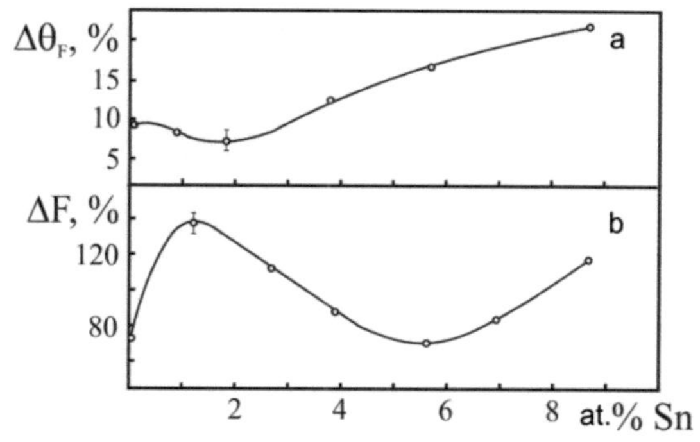

Figure 8. The concentration dependencies of annihilation parameters for Ti-Sn system.

On the basis of the findings related to studying Ti alloys one can state the following. For the deformed Ti alloys the sharply expressed anomalies are typical, which become even stronger with certain concentrations of the alloying elements and when the initial lattice is fully reconstructed as a result of considerable shortening of material interatomic distance. Consequently it is not possible for crystal lattice to preserve initial electron structure. Therefore, probably for the investigated Ti binary alloys we should adopt a concept of autonomy of d-electron matrix subsystem relative to the alloying elements interstitial atoms conduction band, when the wave functions of the matrix atoms d-electrons are overlapped with the wave functions of impurity atoms conduction electrons. The latter is probably correlated somehow with the lattice instability. With this in mind the largest lattice instability are displayed by the investigated Ti alloys systems that contain 1.2 at.% Sn, due to which the maximum change of ΔF annihilation parameter is observed.

2.4. Restoration of structure damages in plastically deformed titanium alloys

In many metals, the structure damages generated at low temperatures are usually "frozen", which enables investigation of their spectrum by measuring some macroscopic properties of the crystal with its subsequent heating. Among the most applicable are the methods of residual electrical resistance measurement, crystalline lattice period measurement, X-ray line profile measurement, etc. For dislocation structure investigation, electron microscopic and neutron diffraction methods were most effective [7, 8].

The main task of investigating metals modified by cold deformation is the differentiation between effects that are related to the presence of the structure defects ensemble in them. These effects can be more or less successfully determined while investigating the processes of recovery and recrystalization of deformed metals by annealing. The most important structure imperfection annealing mechanisms are absorption of point defects by dislocations, mutual destruction of vacancy and interstitial atoms, breakup of point defects aggregations into individual one, etc. As a result of these mechanisms, while heating, the structure damages are partially or fully annealed in different temperature ranges. Eventually the annealing process can be interrupted after recrystallization, due to which full removal of the defect structure and the initial structure restoration is observed.

Of course, in the given temperature range more than one independent or concurrent processes can occur. Then the overall picture of annealing kinetics can be presented as superposition of separate individual processes each of which is responsible for a certain return mechanism. In addition, it appears to be impossible to avoid formation of the complexes, mobility of which can significantly differ from that of the single defects. The annealing kinetics of structure damages in metals, provided that there is one type of defects, can be described by following equation [9, 10]:

$$dP / dt = -\lambda_0 P^n = -A_0 P^n(t) \exp\left[-E_a / kT\right], \tag{10}$$

where $P(t)$ is the change of some material's physical property; T is absolute temperature. Here the activation energy of migration E_a is a typical sign of each of the defects type [11, 12]. If there are several active processes in the crystal then the search of the respective E_a values becomes complicated. At the same time the determination of the physical meaning of each value of E_a also becomes nontrivial. Therefore, it is often assumed that in the given narrow temperature range only one active process is running.

The essence of conducting the isochronal annealing by EPA method is based on the ADAP curve return for a defect material up to annealed state. As a result, the defects annealing states are usually established and the respective activation energy is defined by the following equation [13]:

$$E_a = kT_0 \ln(\nu \cdot k / \alpha E_a), \tag{11}$$

where $\nu \approx 10^{13} s^{-1}$ is a Debye frequency; k=8.62 · 10^{-5}eV/K is the Boltzmann constant; T_0 is average temperature of annealing stage (K); $\alpha = \Delta(T^{-1}) / \Delta t$; $\Delta(T^{-1}) = T_I^{-1} - T_F^{-1}$; T_I and T_F are initial and final temperatures of the stage; Δt is annealing time at a given temperature (T).

The results of annealing investigations for titanium deformed by ε = 50% and 2.7 at.% Zr titanium alloy at different degree of deformation (ε = 5, 10 and 20%) are presented in Fig. 9 as isochronal annealing curves, which are shifted one relative to another by a constant value downward along the Y-coordinate. Isochronal annealing of iodide titanium is implemented in two steps. The first one is a low-temperature stage in the temperature range of 150-350ºC, i.e. the recrystallization temperature threshold for this stage is 150ºC. This stage, which has defects migration energy activation E_{a1} =1.35eV, corresponds to the vacancy complexes. The high-temperature stage is located between 350ºC and 650ºC with activation energy E_{a2} = 1.9eV. It is obvious that in this stage the dislocation defects and the packing defects, as β-phase initiated by plastic deformation, are annealed.

The results of annealing of Ti–2.7 at. % Zr alloy, which was exposed to different degrees of deformation, seems more interesting. It is easy to see that the annealing curves shape for alloy considerably differs from those of Ti that verifies the appropriate role of the alloying element Zr. With an increase of the plastic deformation degree the temperature threshold of recrystal-

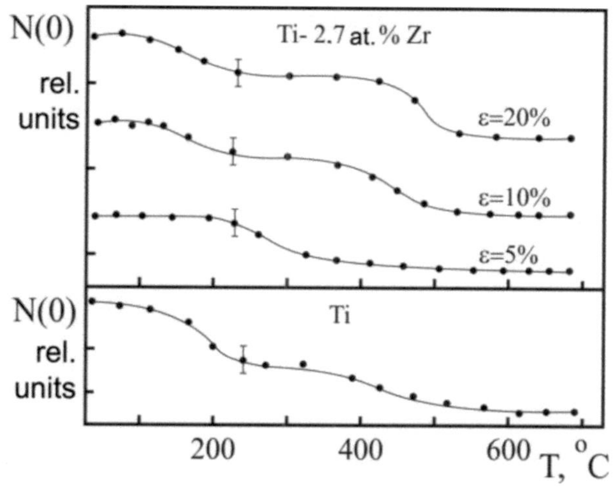

Figure 9. Annealing kinetics of Ti-Zr alloys

lization gradually shifts towards low temperature: from 200ºC at ε = 5% down to 100ºC at ε = 20%. At the same time the I-stage part tends to increase depending on the deformation degree. This effectively confirms successive accumulation of vacancy defects. This stage with E_{a1} = 1.65–1.75 eV, evidently corresponds to vacancy and vacancy-impurity complexes. The II stage of annealing for this alloy tends to narrowing along with an increase of the deformation degree. All narrowing processes of the II stage are also accompanied by an increase of its relative level. Based on these results it can easily be established that it is Zr that serves as the β-phase initiating element. The given high-temperature stage, for which the migration activation energy occupies the interval of E_{a2}=1.85eV, probably corresponds to the dislocation and packing defects annealing with β-phase with split dislocations initiated by plastic deformation.

2.5. The structure modification of Ti-V alloys system as a source of packing defects

The positrons interaction with packing defects is of certain interest because some researchers tend to doubt that positrons can be captured by defects of this type [14]. Therefore, arrangement of special experiments with more precise methods for detecting the latter is the task of high importance. However, in order to ensure with the highest probability packing defects occurrence as a result of plastic deformation, the investigated titanium alloys have to be accordingly chosen. To this effect, the alloying elements must form a complete solubility of solid solution with titanium without eutectic and peritectic in the investigated concentration range of the second component with the successive decrease of the phase transformation temperature. In other words, the alloying element must be β–stabilizer.

Packing defects are a voluminous lesion. The objective of this research was to study packing defects by virtue of comparing X-ray structural analysis results with positron annihilation data. Based on the aforementioned, vanadium (V) was chosen as an alloying element, which is located in the Y-group of the periodic system. The energy of V packing defects formation is ν = 0.1 J/m^2, that is five times greater than for Ti but considerably smaller than for other metals of the transition group. V is a β-stabilizing substitutional element for Ti with the atomic diameter of 2,72Å. The V alloys with 0; 0.5; 1.5; 2.0; 4.0 and 5.8 at. % content were prepared by the technique described above. The plastic deformation by $\varepsilon = 80\,\%$ was implemented by rolling at room temperature. $(10\bar{1}0)$, (0002) and $(10\bar{1}0)$ lines profiles were taken in the filtered CuK_α radiation using the DRON-2 diffractometer. The packing defects formation on prismatic plane $\alpha_{(1010)}$ probability calculation results along with the annihilation parameters are summarized in Table 2.

Materials at. %	State		F	ΔF %	θ_F mrad	$\Delta\theta_F$ %
Ti	annealed	-	0.30	-	6.33	-
	ε=80%	$1.2\ 10^{-3}$	0.44	46	5.83	7.9
Ti-0.5 V	annealed	-	0.28	-	6.36	-
	ε=80%	$4.2\ 10^{-3}$	0.42	50	5.75	9.6
Ti-1.5 V	annealed	-	0.29	-	6.40	-
	ε=80%	$6.2\ 10^{-3}$	0.39	35	5.83	8.9
Ti-2 V	annealed	-	0.23	-	6.40	-
	ε=80%	$7.7\ 10^{-3}$	0.34	49	6.00	6.2
Ti-4.6 V	annealed	-	0.24 0.35	-	6.42	-
	ε=80%	$9.0\ 10^{-3}$		44	6.08	5.3
Ti-5.8 V	annealed	-	0.28 0.35	-	6.29	-
	ε=80%	$10.2\ 10^{-3}$		25	5.79	7.9
Accuracy ±		0.001	0.01	2.0	0.05	0.1

Table 2. The packing defects probability and Ti-V alloys annihilation parameters

One can see that for all investigated alloys the plastic deformation leads to an increase of the parameter *F* by 25-50% with a simultaneous decrease of the Fermi angle θ_F by 5-10%. The changes of the annihilation parameters depending on the composition are of monotonous nature. The packing defects probability on the basal plane both for Ti and alloys remains practically without changes and equals $\alpha_{(0001)} = 2\ 10^{-3}$, whereas on the prismatic plane (1010) the probability monotonically increases from $4.2\ 10^{-3}$ up to $10.2\ 10^{-3}$ depending on V concentration. Thus, the packing defects formation under plastic deformation in Ti – V alloys became an established fact and alloying with V only facilitates this.

The results of the isochronal annealing for deformed Ti, V and Ti – V alloys are presented in Fig. 10.

On the basis of stage II annealing results analysis one can notice that among the investigated materials the packing defects are most pronounced only in the alloy with 2 at.% V, which corresponds to the temperature range 350 - 720ºC with $E_{a2} = 2.35$eV. For all other alloys this

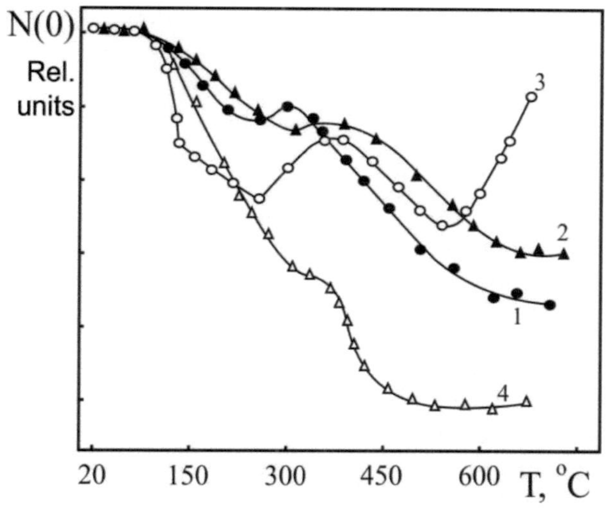

Figure 10. The concentration dependencies of annealing kinetics for deformed Ti-V alloys 1 – Ti; 2 – Ti – 2.0 at. %V; 3 – Ti – 4.6 at.%V; 4 – V

process is shaded by the bound vacancy-impurity complexes, which are caused by plastic deformation, and the impurity atoms atmosphere formation around packing defects.

3. Radiation modificationof the titanium alloys properties

3.1. Problem statement

The study of positrons behavior in the plastically deformed metals showed high sensitivity and selectivity of the EPA method to the structure damages in these materials. Therefore it is natural that investigators tend to use this method to learn about radiation effects in solids as a result of nuclear irradiations of a material. This irradiation is accompanied by a number of new phenomena. The most important among them are nuclear reactions and related to them change in the elemental composition, point defects formation and crystal integrity disturbance, point defects aggregations occurrence and matrix disturbance caused by atomic collisions cascades, etc.

It is clear that without careful and detailed study of all aspects of nuclear radiation interaction with material and its consequences it is impossible to predict behavior of the materials in the field of strong ionizing radiation. The positron annihilation methods are promising and sufficiently informative for investigations of this kind.

As known, interaction of nuclear radiation with a material occurs by elastic and inelastic collisions channels. It is impossible to trace the process of radiation damage, which happens during 10^{-13}–10^{-11}s. Therefore, using different experimental methods the final structure of radiation damaged material is usually studied, which is in the state of equilibrium with an environment. Consequently, investigation and control of construction materials radiation damageability is the task of primary importance and, undoubtedly, attracts a considerable scientific and practical interest. Though, as of today there are practically no systematic, detailed and purposeful investigations of titanium and its alloys radiation properties. In this chapter the results of authors' own investigations in this field are thoroughly described.

3.2. The methodology of materials irradiation on accelerator and reactors

Reliable and reproducible results of the investigation of ionizing radiation influence on the solid can be obtained only under conditions of guaranteed high accuracy of measurement of irradiated target temperature. The final structure of the material is determined by the conditions of irradiation.

Since positron annihilation methods are mainly sensible to vacancy defects, in this case the problem was to preserve just this type of defects during the irradiation process of the investigated material. Taking into account these circumstances, the irradiation of the samples on the electron accelerator ELU-6 and isochronous accelerator U-150 was conducted in the air atmosphere with the water-cooled base and forced blow-off of the sample with liquid nitrogen vapor. With the charged particles intensity of $(1.5 - 2) \times 10^{16}$ m^{-2} s^{-1}, the sample's temperature did not exceed 60–70^{0}C.

The major portion of the reactor irradiation, related to investigation of neutron flux influence on metals, was implemented using the nuclear reactor VVR-K at the National Nuclear Centre of the Institute of Nuclear Physics of the Republic of Kazakhstan. The reactor's nominal power is 10MW. Energy distributions of the thermal and fast neutrons fluxes for irradiated channels were determined by the activation analysis method. For thermal neutrons cutoff the method of samples screening by cadmium was applied. After irradiation the materials were exposed to the chain of "hot chambers" where they undergo cutting and dosimetry control. The samples temperature in the irradiation process was taken to be equal to the temperature of the primary-coolant system heat carrier (+80ºC).

3.3. Titanium structure modification as a result of electrons irradiation

As any other charged particles, while interacting with the crystalline lattice, the high energy electrons experience losses of energy on excitation, ionization and atoms displacement. For metals the first two electron interaction processes usually end without consequences. The consequences of elastic interactions depend on electron and recoil atom mass ratio as well as recoil energy E_p. If the recoil energy is greater than the defect formation threshold energy ($E_P > E_d$), then atom will leave its place in the lattice, which leads to formation of the elemental Frenkel pair, i.e. interstitial atom and vacancy. When E_p values are high the displacement cascades can appear and they consist of two or three vacancies and a certain number of interstitial atoms. The latter move towards the sinks or recombine with vacancies at the room temperature. Therefore, as a consequence of high energy electrons irradiation vacancy defect structure is generally formed in the crystal, which can be effectively studied by positrons diagnostics methods.

To this end, iodide titanium samples of high purity were irradiated by E = 4MeV electrons with intensity of about 5 10^{12} $cm^{-2}s^{-1}$ and up to 3.7 10^{17}; 10^{18}; 3.7 10^{18}; 10^{19} and 3.7 10^{19} cm^{-2} fluencies

http://dx.doi.org/10.5772/55485

at a temperature lower than 70ºC with the following ADAP spectra measurement and structure sensitive annihilation parameters determination. The results of these investigations are shown in Table 3. One can see that the positrons and conduction electrons annihilation relative probability practically grows in linear fashion with the fluence increase and rate of this growth decreases only at the last two fluence values: 10^{19} and 3.7 10^{19} cm^{-2}. At the same time the Fermi momentum angle θ_F has a trend to stabilization at 5.70 mrad.

Fluence, cm^{-2}	*F*	θ_F, mrad.	R_C	E_a, eV
0	0.21	6.33	-	
$3{,}7\cdot10^{17}$	0.36	5.85	1.52	
10^{18}	0.40	5.75	1.55	1.23
$3{,}7\cdot10^{18}$	0.45	5.70	1.62	1.28
10^{19}	0.49	5.72	1.58	
$3{,}7\cdot10^{19}$	0.53	5.69	1.61	
Accuracy ±	0.01	0.02	0.05	0.05

Table 3. The annihilation parameters dose dependence for titanium irradiated by electrons

If the positrons and conduction electrons annihilation probability growth with a fluence increase can indicate an increase in the respective point defects concentration, then the Fermi momentum practical constancy indicates lack of changes in the electron structure of the latter. In other words, the vacancy defects configuration on the reached level of fluence remains the same. On the basis of positrons capture model one can calculate the average size of the defect region created in Ti as a result of electron irradiation: $r = 0.81$Å. This means that the positrons capture centers in this case are really the vacancies.

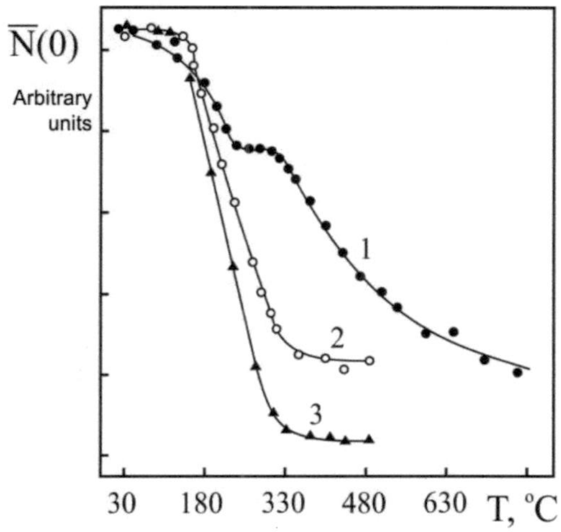

Figure 11. The structure damages annealing in deformed and electron irradiated titanium 1- deformed on ε =50%; 2- electron irradiated $\Phi_1 = 10^{18}$ cm^{-2}; 3 - electron irradiated $\Phi_2 = 10^{19}cm^{-2}$.

This statement is also confirmed by calculation results of the configuration parameter R_c, which is determined according to the positrons capture model (Table 2). It can be seen that within the calculation error R_c remains constant ($R_c = 1.55\pm0.05$), that is regardless of electrons fluence the radiation defects configuration remains invariant. Consequently, the observed increase of the probability of positrons annihilation with the conduction electrons F is caused only by a respective increase of radiation defects in Ti. These data are verified by the isochronal annealing results performed for three cases (Fig.11). One can see that in the temperature interval 170-240ºC only one return stage for irradiated materials is observed, which is related to removal of radiation defects regardless of electrons fluence. The higher return effect value for $\Phi_2=10^{19}$ cm^{-2} fluence in comparison with $\Phi_1=10^{18}$ cm^{-2} fluence also confirms enhanced concentration of vacancy defects formation. The defect migration activation energy value did not exceed $E_a = 1.22\pm0.05$ eV and according to the data of [11, 12] it enabled us to identify them as point defects. Thus, the titanium structure modification by high energy electrons irradiation leads to formation of point defects of vacancy type, the concentration of which depends on the irradiation fluence.

3.4. Radiation induced modification of titanium structure at the helium ions irradiation

Use of accelerators of high energy charged particles plays quite an important role in studying radiation modification fundamental problems. This is first of all important for prediction of the construction materials behavior and change of their properties. To this end, the structure modification process of titanium binary alloys that contain 0; 1.2; 2.5; 3.3; and 4.1 at. % Ge; 1.2; 2.5; 4.3; 6.2 and 7.6 at.% Sn and 1.4; 2.9; 5.1 and 10.3 at.% In was performed. The modification was realized by α-particles irradiation with $E = 50$ MeV with the flux density 1.5 10^{12} cm^{-2} c^{-1} and the temperature not higher than 70ºC.

It should be noted that irradiation by α-particles with E=50 MeV causes significant deformation of the spectra shape that considerably exceeds the influence of the plastic deformation of sufficiently high degree ε =50% (Fig.11). The probability of positron annihilation with conduction electrons as a result of irradiation by α-particles in all cases significantly exceeds its values for alloys both in initial state and also after strong plastic deformation with simultaneous and significant decrease of the Fermi angle θ_F. It is important that the nature of change of the concentration dependence of these parameters is preserved from one alloy to another. Meanwhile, those alloys, in which abnormally high increase of the annihilation parameter after plastic deformation is observed, also show this tendency after α-particles irradiation. For some alloys these changes exceed by more than twice the corresponding indicators for the deformed state. This indicates that the α-particles irradiation appears to be more effective on structure changes in metals than plastic deformation.

Stability of the each alloys system towards the α-particles exposure depends on the nature and concentration of the alloying element. The smallest stability towards the α-particles exposure was observed in the alloys containing 0.8 at.% Ge; 1.2 and 7.6 at.% Sn, as well as 1.4 and 7.4 at. % In. Therefore, the damageability of alloys of the indicated compositions in relation with ? Ti under α-particles irradiation is higher than for alloys of other compositions (Fig.12).

http://dx.doi.org/10.5772/55485

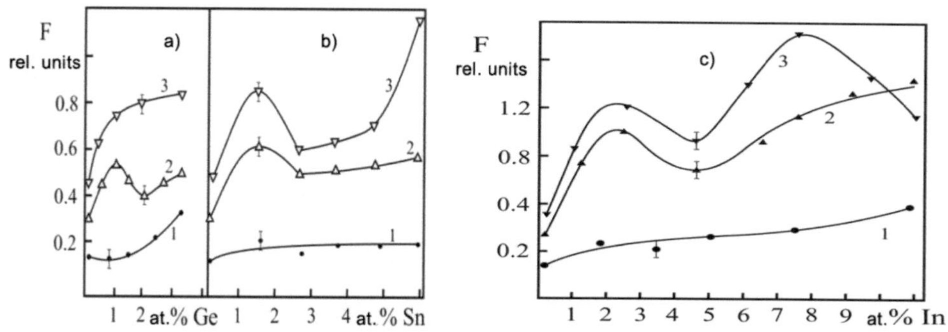

Figure 12. The alloys structure modification: Ti-Ge (a), Ti-Sn (b) and Ti-In (c) in different states: 1 – annealed; 2 – deformed by ε=50 %; 3 – irradiated by α-particles with E=50MeV

On the basis of positrons capture model and assuming formation of new allocations in crystal structure due to α - particles irradiation we can estimate the average size of these regions. The sizes are: r = 16 Å for Ti–1.4 Ge at.% alloy; r = 15Å for Ti-1.4 at.% In alloy and r= 20 Å for Ti – 7.6 at.% Sn alloy. These data are confirmed by the isochronous annealing results.

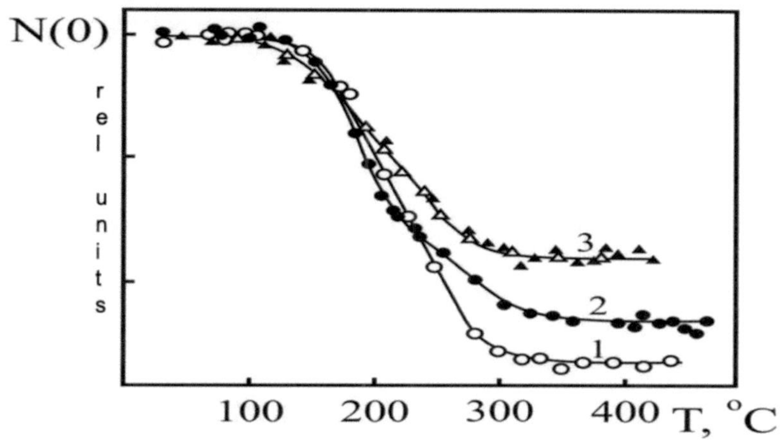

Figure 13. Annealing kinetics of titanium alloys, irradiated by α - particles with E = 50MeV: 1 - 1.2 at.% Ge 2 - 1.2 at.% Sn 3 - 2.9 at.% In

The results of isochronal annealing of the Ti irradiated alloys presented in Fig.13 reveal only one stage of return regardless of the composition of the alloy. This annealing is completed at the temperature of 300-320ºC with radiation defects migration activation energy of E_a=1.50-1.55eV and corresponds to vacancy complexes that occur in the Ti alloys' α–phase.

3.5. The dose dependence of titanium alloys structure modification under α–particles irradiation

This characteristics is estimated by radiation defects accumulation kinetics at α–particles irradiation with 10^{14}; 3.2 10^{14}; 3.2 10^{15} and 10^{16} cm^{-2} fluences on the example of the Ti-Ge alloys system. The α–particles energy was E=29 MeV with the beam intensity 1.5 10^{12} $cm^{-2}c^{-1}$ (Fig. 14). One can see that the accumulation curve character is practically not dependent on the Ge concentration. The 10^{14}÷5 10^{15} cm^{-2} fluence is correspond to the incubation period of radiation defects accumulation. Further, the defects accumulation obeys the point defects clusterization principle.

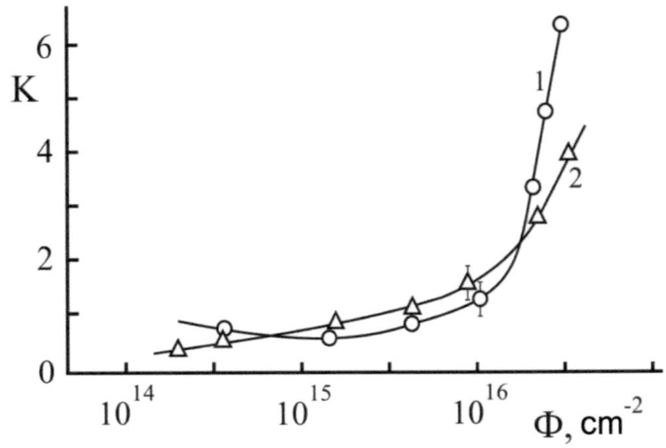

Figure 14. Positrons capture efficiency dosage dependencies for Ti (1) and Ti–3.1 at.% Ge (2)

3.6. The peculiarities of structure modification of titanium alloys irradiated by protons

By the time of setting the experiment for the purpose of studying radiation modifications of Ti and its binary alloys structure under irradiation by high energy protons, there was no a single research work with published results that was devoted to this problem. Therefore, investigation of radiation damageability caused by strong protons beam was realized on the example of Ti binary alloys, alloyed by Sn in aforementioned concentration. The Ti–Sn alloy samples in the initial annealed and deformed ($\varepsilon = 50\%$) states underwent irradiation by protons with $E = 30$ MeV up to two-value (5 10^{15} and 2.5 10^{16} cm^{-2}) fluence for the purpose of elucidating not only the dopant agent role but also the material's history in the formation of the final defective structure of alloys. It is necessary to point that for the protons with $E = 30$ MeV, the thickness of the samples used (1 mm) was absolutely insufficient for providing their complete deceleration. The calculated value of protons energy on the backside of the samples differed from protons initial energy only by 5–6MeV [15, 16]. Therefore, all investigated samples were irradiated actually by shooting. The results of this investigation are presented in Table 4. One can see that for the materials' initial state the alloying elements concentration increase has a

Alloys composition, at.%.	Materials state					
			Protons fluence, E=30MeV			
			$5{\cdot}10^{15}$ cm $^{-2}$		$2.5{\cdot}10^{16}$ cm $^{-2}$	
	Annealed	Plastic deformed *ε=50%*	After annealing	After deformation *ε=50%*	After annealing	After deformation *ε=50%*
Ti	0.22	0.38	0.30	0.41	0.42	0.45
Ti-1.2 Sn	0.25	0.44	0.36	0.32	0.38	0.36
Ti-2.5 Sn	0.22	0.39	0.29	0.34	0.31	0.38
Ti-4.3 Sn	0.26	0.41	0.28	0.37	0.38	0.40
Ti-6.2 Sn	0.27	0.41	0.34	0.33	0.37	0.46
Ti-7.6 Sn	0.26	0.43	0.30	0.36	0.41	0.41
Accuracy ±	0.02	0.02	0.02	0.02	0.02	0.02

Table 4. The positrons annihilation probability in titanium alloys irradiated by protons

weak influence on positrons annihilation process nature both in respect of the probability $W_P = S_p/S_o$ and the Fermi angle θ_F.

At the same time the proton irradiation of annealed alloys leads to a significant increase of positrons annihilation probability at a fluence of 5 10^{15} cm^{-2}: by 36% for Ti and by 50% - on an average for alloys containing 0.8 and 1.5 at.% of Ge. If we take the W_P value for the alloys deformed by $\varepsilon = 50\%$ as saturating, then the results for alloys irradiated by protons from an annealed state confirms the absence of tendency of the annihilation parameter to saturate within the range of reached fluence level. The annihilation parameter increase with the fluence characterizes the respective increase of radiation defects concentration in the materials' structure. The largest increase of positron traps as a result of proton irradiation is observed in the alloys containing 1.2 and 7.6 at.% Sn, i.e. these alloys, as in the case of α–irradiation influence, manifest certain instability to proton irradiation influence.

Analysis of the results of alloys irradiation to up to $5{\times}10^{14}$ cm^{-2} fluence in the previously deformed state testifies a completely opposite picture. In this case the positron annihilation probability takes substantially smaller values than before irradiation practically for all investigated alloys of this system. This tendency remains even after re-irradiation of up to $2.5{\times}10^{16}$ cm^{-2} fluence, which indicates the significant role of the material's history and the dopant agent nature in the formation of the structure damages caused by proton irradiation. The primary decrease in the positron annihilation probability W_P caused by proton irradiation of deformed materials is probably related to the appropriate decrease in the efficiency of structure damages towards positron trapping. The following increase of W_P is obviously caused by probable radiation-stimulated reconstruction of alloys dislocation structure as a result of proton irradiation.

The calculation of the average size of these centers on the basis of a one-trap model positron capture for Ti–2.5 at% Sn alloy irradiated by E = 30 MeV protons up to 2.5 10^{16} cm^{-2} fluence gives R_V = 10Å, i.e. in the investigated materials one can assume a generation of vacancy clusters.

These very peculiarities of the concerned problem are well confirmed by the results of isochronal annealing of structure defects. As an example, Ti and Ti–7.6 at% Sn alloy were chosen for studying annealing. These samples underwent all of the three abovementioned types of external exposure. The results of these investigations are reflected in Fig. 15(a) and (b) as curves of isochronal annealing. In each case, in order to establish the nature of the structural transformations, the curves of annealing for the deformed materials obtained earlier are provided. Comparing the annealing results for all states of the materials is useful as it enables formulation of quite important conclusions about some redistribution of defects in the crystal structure of materials that underwent a combined treatment. One can observe a pronounced low-temperature stage for Ti caused by an irradiation by protons up to 2.5 10^{16} cm^{-2} fluence from a deformed state within the temperature range of 60 – 220°C (Fig. 15(a), curve 2). It occurred as a result of the transformation, evolution and redistribution of the initial defect structure generated by an intense plastic deformation under a heavy proton radiation. This stage significantly differs from that of annealing curve for deformed titanium (curve 1) both by form and by temperature region of manifestation.

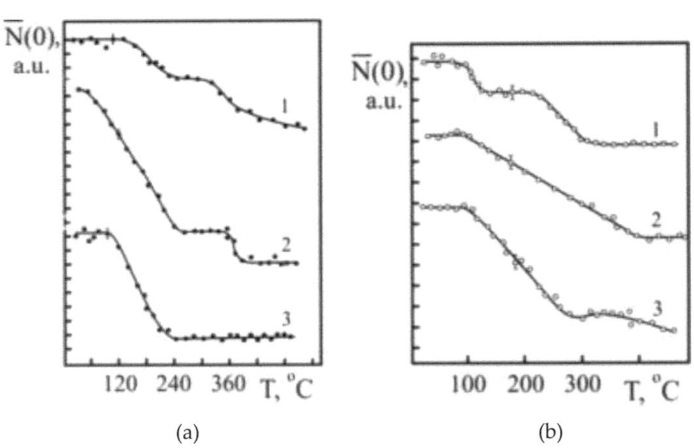

Figure 15. Kinetics of Ti (a) and Ti–7.6 at% Sn alloy (b) annealing under different types of exposure. 1 – deformed by ε = 50%; 2 – irradiated by protons in a deformed state; and 3 – the same in an annealed state.

In addition, one can observe a second high-temperature stage with ΔN_2 = 2.5% in the temperature range of 330–360°C. The defect migration activation energies by stages were equal to E_{a1} = 1.21 eV and E_{a2} = 1.93 eV, respectively. This fact confirms an assumption about dislocation structure evolution and its partial transformation into the vacancy structure.

3.7. The metals and their alloys structure modification under neutron irradiation

Relatively high stability of some Ti–Al system alloys to fast α–particle influence was shown above experimentally. Though, in the literature one can encounter contradictory assertions about radiation characteristics of these alloys [16]. For tackling this problem alloys of this system of compositions investigated earlier underwent irradiation by fission neutrons.

Irradiation of the annealed materials by fission neutrons with E>0.1 MeV was conducted at the fluence up to 2 10^{22} cm^{-2} according to the method described above. The results of the realized investigations are presented in Fig.16.

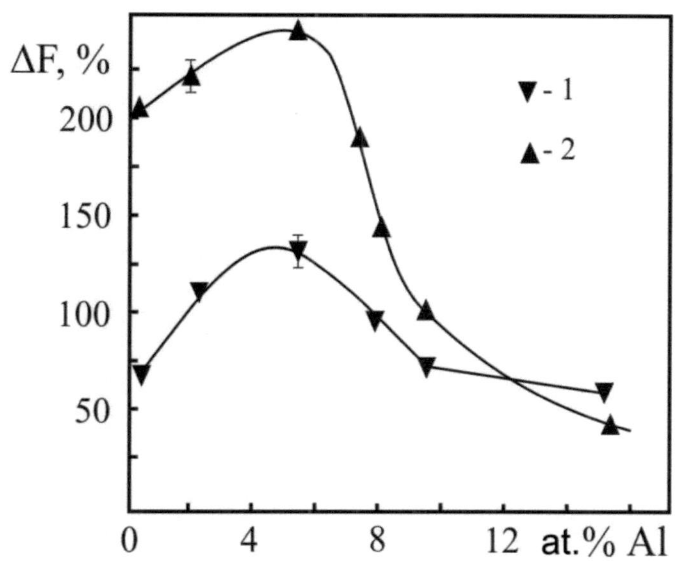

Figure 16. The concentration dependencies of annihilation relative probability for Ti-Al alloys, are subjected to plastic deformation (1) and neutron irradiation (2)

In the titanium alloys that are inclined to phase transformations, as a result of plastic deformation one should expect formation of defects related to the $\alpha \rightarrow \beta$ transformation. The vacancy-dislocation structure, which appears after strong neutron flux, can stimulate formation of certain allocations in the crystal, an integral part of which is the lattice instability related to the energy excess introduced by irradiation. One can see from the data that for Ti and 5.2 at.% of Al alloy the annihilation parameter ΔF increases under the neutron irradiation by more than three times, while after plastic deformation by $\varepsilon = 50\%$ it was only 64 and 127 %, respectively. Such behavior of the annihilation parameter indicates that with an increase of Al content in the alloy considerable changes in the defects electron density distribution occur.

At the same time sufficiently close values of the annihilation parameter for heavily doped alloys with 10.2 and 16.5 at.% of Al as a result of deformation and neutron irradiation can validate the possibility of structure damages formation in these alloys that assures an equivalent efficiency of positrons capture potentials. Materials, which manifest this regularity, possess enhanced stability to external influence, including a neutron irradiation. The Ti alloys containing more than 10 at.% of Al probably belong to this category of materials.

For investigation of the radiation defects accumulation kinetics, alloys of Ti-Al system were irradiated by fission neutrons in the wide range of fluencies: 10^{15}; 10^{16}; 10^{17}; 10^{18}; 10^{19} and 2 10^{19} cm^{-2}. Fig 17a depicts the dose dependence of ΔF parameter for Ti and its three alloys with Al. The sharply distinct nature of change of this parameter related to the alloys composition should be noted. Given these conditions one can assume that in the alloys, which are different by composition, the changes of the defects concentrations and their configuration can also be different. The strongly concentrated alloys are more stable to neutron exposure and for this reason the fluence increase of the latter is not accompanied by sharp changes of the annihilation parameters and this is probably stimulated by an interatomic bond energy increase in these alloys comparing with Ti and Ti–5.2 at.% Al alloy.

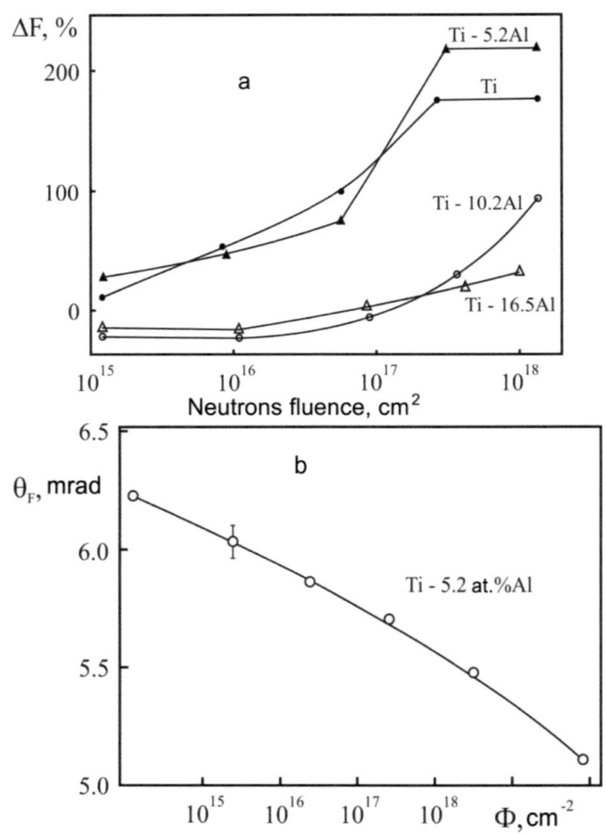

Figure 17. The dosage dependence of annihilation parameters change kinetics in Ti-Al alloys irradiated by fission neutrons (a) and Fermi momentum (b).

http://dx.doi.org/10.5772/55485

The certain clarity about the processes can be obtained from the annealing data. Firstly, let us consider the radiation defects annealing spectra in Ti–5.2 at.% of Al alloy, irradiated by neutrons at different fluencies (Fig.18a). As a result of neutron irradiation to up to 10^{17} cm^{-2} fluence, the generated radiation defects in the crystal structure are annealed in one stage in temperature range of 110 - 210ºC (curve 1) with migration energy activation E_{a1} = 1.27 eV. This stage evidently corresponds to the return of small vacancy clusters.

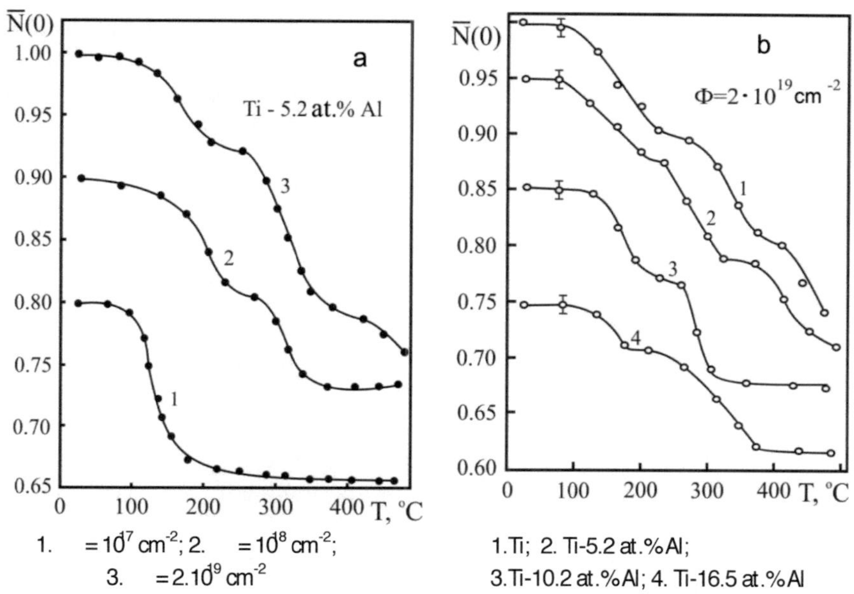

Figure 18. Dosage (a) and concentration (b) dependencies of annealing kinetics for Ti-Al alloys irradiated by neutrons

The fluence increase by single-order (to up to 10^{18} cm^{-2}) leads to the radiation defects occurrence in the structure and these defects are annealed in two stages (curve 2) and with the fluence of 2 10^{19} cm^{-2} the full healing of the structure damages is performed in three evidently expressed annealing stages (curve 3).

A considerable interest is presented by the annealed defects spectra against the alloys composition irradiated at the same neutron dose 2 10^{19} cm^{-2}. It is easy to determine that in the different alloys the radiation defects accumulation process occurs differently (Fig.18b).

In conclusion, a comparative analysis of titanium radiation damageability under the four types of particle radiation influence can be performed: electrons, protons, α-particles and fission neutrons. This comparison can only be approximate, since it is practically impossible to ensure same conditions for all cases. The titanium and its alloys radiation damageability is substantially higher at α-particles irradiation. Taking into account radiation defects generation rate the most damaging are (detrimental) α-particles, which are then followed by fission neutrons, protons, and electrons.

4. Titanium alloys defect structure modification by hydrogen saturation method

It is generally acknowledged that alterations in the metal electron structure caused by hydrogen should be taken into account when interpreting physical properties of the metal-hydrogen system. This is especially important for the Ti–H system, because of the high absorption capacity of titanium with respect to hydrogen, which to a significant degree have an influence on its technological properties. According to the well-known proton model, the proton gas penetrates into the electron shells and changes their energy state. At the same time the intensity of the force fields grows with an increase of the temperature of the system, which is accompanied by intensification of interaction between proton gas and electron shells.

In order to tackle this task, as an objects of the experimental study iodide titanium and a system of Ti–Al alloys (up to 50 at.% Al) were taken [13,17]. For hydrogenation, the original materials were annealed at 900^0 C, deformed by rolling by ε =50%, and irradiated by 50MeV α-particles up to a fluence of 5 10^{15} cm^{-2}. Hydrogenation was implemented by Siverts saturation method at the temperature of 200^0 C during 3 hours and under 500^0 C during 1 hour. The hydrogen pressure was (4.9–5.9) 10^5 Pa. Hydrogen was produced by desorption of the hydrides $LaNi_5H_X$ and TiH_X. Before hydrogenation, the samples were kept under the room temperature in vacuum of 0.13–1.33 Pa for 10 hours directly in the reactor, when hydrogen was admitted after the previous processing. The pressure was measured by a standard manometer with accuracy of 300 Pa with the volume of the reactor system being equal to 4 × 10^{-4} m^3. After samples hydrogen saturation at T=200ºC in any different initial state the change of their weight could not be locked. At T=500ºC the hydrogen saturation flows more intensively due to hydrogen diffusion acceleration, and samples' weight substantially increases. For interpretation of the investigation results the annihilation parameters shown in Table 5 were used.

Material state	Titanium			Ti-5.2 at. % Al			Ti-1.4 at. % V		
	F	**θ_F mrad**	**ΔF %**	**F**	**θ_F mrad**	**ΔF %**	**F**	**θ_F mrad**	**ΔF %**
Annealed (before hydrogenation)	0.26	5.82	-	0.34	5.61	-	0.27	6.12	-
Annealed+H (200°C)	0.29	5.85	15	0.33	5.83	0	0.29	5.92	9
Annealed+H (500°C)	0.37	5.92	41	0.46	5.92	36	0.33	5.91	23
ε **50%**+H (200°C)	0.32	6.05	22	0.46	5.82	35	0.38	6.22	43
α+H (200°C)	0.47	5.18	81	0.48	5.33	42	0.46	5.03	72
α+H (500°C)	0.34	6.03	28	0.45	5.61	32	0.39	5.63	45
Ti pressed powder	0.40	5.25	54		Error ±		0.01	0.05	1.0
TiH*x* Pressed powder	0.50	5.00	90						

Table 5. Annihilation parameters of hydrogenated titanium alloys

In ideal defect-free single crystals hydrogen dissolves in the metal, occupying the lattice interstitial positions and causing displacement of the metal atoms from their equilibrium positions. In the real crystals, though hydrogen segregates in various lattice defects, which leads to reduction of probability of positrons capture. But hydrogen saturation of materials in annealed state at T = 200ºC does not practically affect annihilation parameters. If we increase the temperature of the hydrogen saturation process to up to T=500ºC, then hydrogen absorption accelerates, which consequently causes more active hydrogenation of that material. This is effectively reflected in the respective increase of F probability, but angle of Fermi momentum θ_F remains constant in the range of calculation error. This means that the hydrogen saturation process is not accompanied by new defects formation in the metal's structure and electron subsystem in positrons locating places does not experience significant reconstruction.

Plastically deformed metals facilitate accumulation of a considerable amount of hydrogen. The hydrogen saturation at comparatively low temperature (200ºC) is obviously accompanied by atomic hydrogen capture by structural damages. The new complexes appear in crystalline lattice which decrease the capture efficiency of positrons localization centers, previously introduced by the plastic deformation. The hydrogen saturation temperature increase up to 500ºC converts materials into a hydride state, which caused the cast, compact metals to crumble into powder. The reason for destruction of the compact metal after hydrogen saturation under 500ºC can be formation of cracks of deformative nature related to the hydrides formation. The penetrated hydrogen is dissociated in the internal surface of these defects but with formation of less mobile molecules which with more intensive arrival from outside gradually become bigger by volume and eventually cause sample's destruction.

In order to establish the nature of the observed phenomena, the hydrogen accumulation nature in Ti and its alloys with Al in annealed state was investigated. In clean Ti, there are practically no reasons that could prevent hydrogen from accumulation to significant concentrations (Fig. 19a). At the same time, intensity of hydrogen absorption by annealed alloys of Ti-Al system sharply drops with an increase of Al concentration in the alloys (Fig.19b). Therefore, the capabilities of Al, as an absorber of hydrogen, are rather limited comparing to those of Ti. This is reflected in the annihilation parameter changes.

Thus, the observed changes in the electron structure of the defect material indicate, on the one hand, on hydrogen's significant role in its formation and, on the other hand, on enhancement of the interaction of hydrogen with the material due to the presence of damages of deformative and radiative character.

Isochronal annealing behavior of these materials is presented in Fig.19. For irradiated by electrons and not saturated by hydrogen Ti one stage of restoration is allocated under annealing in the temperature range 170-240ºC with E_a= 1.22 eV. As a result of irradiation of the hydrogen saturated metal, the displacement of the starting point of the first stage recovery up to 225- 230ºC is observed, which finishes near 330ºC. Therefore, the bound state vacancy-hydrogen in Ti has a higher temperature range of dissociation and annealing with E_a = 1.38 eV, than simple vacancy defects.

Fig. 19 Hydrogen accumulation kinetics (a), Al influence on hydrogen accumulation rate in Ti (b) and defects annealing kinetics in titanium (c)

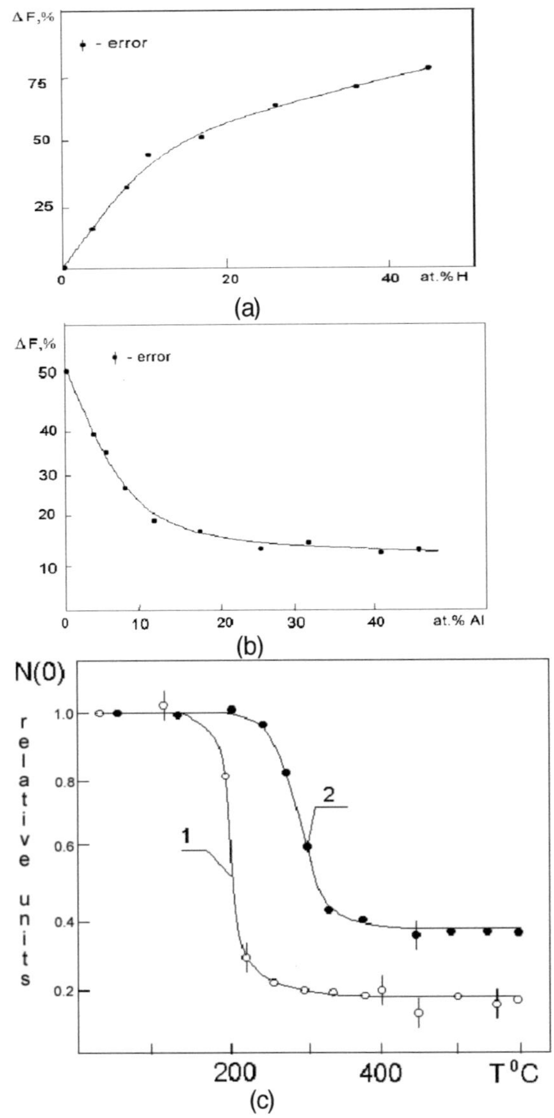

Figure 19. a. Kinetics of hydrogen accumulation *ΔF* in titanium. b. Effect of aluminum addition on hydrogen accumulation *ΔF* in titanium.c. Annealing kinetics of radiation defects in titanium

5. Conclusion

Complex and systematic investigations of the electron structure of the titanium binary alloys depending on the type and concentration of the alloying elements and the modification degree by plastic deformation method enabled formulation of the following conclusions:

- for the initial state the alloys electron structure reveals a weak dependence on type and concentration of an alloying element, whereas the structure modification by plastic deformation causes nonmonotonic changes to the annihilation parameters. At the same time, the main changes in the character of the annihilation parameters are observed at deformations of up to $\varepsilon \leq 30$ %, above which saturation is observed;
- it was shown by X-ray investigations that in the metals modified by deformation one can find a certain amount of new phase precipitates which are interpreted as the β - phase of base metal in the matrix of α-phase, with vacancy-dislocation structure and electron density substantially smaller than in the initial phase;
- it was shown that after electron irradiations, mono vacancies emerge in the titanium crystalline lattice and they remain at the room temperature; these mono vacancies ensure capture of positrons, electron structure and configuration of which do not depend on electrons fluence and significantly differ from the corresponding characteristics of vacancy formations generated by plastic deformation;
- it was established that the electron structure and stability of each alloys system to the influence of high energy α-particles are the function of alloying elements nature and their concentration;
- the possibility of the irradiation-induced swelling suppression in the titanium alloys by selective alloying and preliminary structure defects introduction was experimentally demonstrated;
- the role of the initial structure defects in titanium alloys under the high energy proton irradiation is manifested in their transformation, evolution and redistribution with sharply distinct electron structure;
- it was shown that in the deformed state hydrogen accumulation in titanium occurs in the defect regions with hydride formation which afterwards leads to sample destruction;
- for hydrogen corrosion reduction it is necessary to use titanium alloyed by Al (over 5.2 at. %);
- the radiation effects in preliminary hydrogen saturated titanium manifested themselves in emergence of hydrogen atom-vacancy coupled state.

Author details

Kanat M. Mukashev[1] and Farid F. Umarov[2]

1 Department of the Theoretical and Experimental Physics of the National Pedagogical Unversity after Abai, Almaty, Kazakhstan

2 Department of Geology and Earth Physics of the Kazakh-British Technical University, Almaty, Kazakhstan

References

[1] Li-witing, G. Thermalization of Positrons in Metals. In book: Positrons annihilation in solids. (1960). pp., 17-20.

[2] Holland, A. Current investigations of point defects in metals. In book: Point Defects in Solids. Mir.(1979). pp., 243-370.

[3] Stewart, A. T. Positron annihilation in metals. Canad. Jour. Phys. (1957). , 35, 168-201.

[4] Mantl, S, & Triftshauser, W. Defect annealing studies of metals by positron annihilation and electrical resistiving measurements. Phys. Rev. (1978). N4,-, 17, 1645-1652.

[5] Regnier, P, & Dupony, I. M. Prismatic slip in Be and the relative ease of qlade in H.P.C. metals. Phys.Stat.Sol. (a). (1970). N1.-, 39, 79-93.

[6] Miller, G. L. Zirconium. In.Lit. (1955). P.320.

[7] Klebro, L. M, Hargriva, M. E, & Loretto, M. X. The internal energy change at metals return and recrystalization. M.: Metallurgy. (1966). pp., 69-122.

[8] Larikov, L. N. Plastic deformation and atoms mobility in crystalline lattice. In book: Metals, Electrons, Lattice. Kiev, Naukova Dumka. (1975). pp., 315-354.

[9] Orlov, A. N. Trushin Yu.V. The energy of point defects in metals. Energoatomizdat. (1983). P.57.

[10] Ballufi, R. M, Keller, J. S, & Simmons, R. O. The modern state of knowledge about point defects in metals with FCC-Lattice./ In book: The metals return and recrystalization. Metallurgy. (1966). pp., 9-68.

[11] Mikhalenkov, V. S. To question about plastic deformation influence on positrons annihilation in metals. Ukr. Phys. Journ. (1972). B.17. N5. pp., 840-841.

[12] Mikhalenkov, V. S. Recovery of the angular distributionof annihilation photons by annealing of deformed copper allos. Phys. Stat. Solid. (a) (1974). N2.-, 24, k111-k113.

[13] Mukashev, K. M. A slow positrons physics and positron spectroscopy.- Almaty. (2010). P.508.

[14] Shishmakov, A. S, & Mirzaev, D. A. Khmelinin Yu. F. The X-ray diffraction on packing defects in metals with HCP-lattice. Met. Physics and Metallurgy. (1974). B.37. N2. pp., 313-321.

[15] Mukashev, K. M, & Umarov, F. F. Positron annihilation in titanium alloys modified by proton irradiation. Radiation Effects & Defects in Solids. pp.1-11,167(1)(2012).

[16] Dextyar, I. Ya., Mukashev K.M., Chursin G.P. The high energy α-particles irradiation influence on electron-positron annihilation in titanium alloys. Voprosy atomnoy nau-

ki i techniki. Ser. Fizika radiat. Povrejdeniy I radiat materialovedenie. (1982). N4 (23)-12-16pp.

[17] Mukashev, K. M, & Umarov, F. F. Hydrogen behavior in electron-irradiated titanium alloys studied by positron annihilation method. Radiation Effects & Defects in Solids pp.415-423, 162(6) (2007).